Advances in
IMMUNOLOGY

VOLUME **100**

Immunopathogenesis of
Type 1 Diabetes Mellitus

ASSOCIATE EDITORS

K. FRANK AUSTEN
Harvard Medical School, Boston, Massachusetts, USA

TASUKU HONJO
Kyoto University, Kyoto, Japan

FRITZ MELCHERS
University of Basel, Basel, Switzerland

JONATHAN W. UHR
University of Texas, Dallas, Texas, USA

EMIL R. UNANUE
Washington University, St. Louis, Missouri, USA

Advances in IMMUNOLOGY

VOLUME **100**

Immunopathogenesis of Type 1 Diabetes Mellitus

Edited by
EMIL R. UNANUE
Washington University, School of Medicine, St. Louis, Missouri, USA

HUGH O. McDEVITT
Stanford University, School of Medicine, Palo Alto, California, USA

Series Editor
FREDERICK W. ALT
Howard Hughes Medical Institute, Boston, Massachusetts, USA

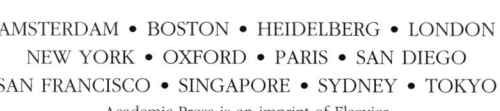

AMSTERDAM • BOSTON • HEIDELBERG • LONDON
NEW YORK • OXFORD • PARIS • SAN DIEGO
SAN FRANCISCO • SINGAPORE • SYDNEY • TOKYO
Academic Press is an imprint of Elsevier

Academic Press is an imprint of Elsevier
32 Jamestown Road, London, NW1 7BY, UK
Radarweg 29, PO Box 211, 1000 AE Amsterdam, The Netherlands
Linacre House, Jordan Hill, Oxford OX2 8DP, UK
30 Corporate Drive, Suite 400, Burlington, MA 01803, USA
525 B Street, Suite 1900, San Diego, CA 92101-4495, USA

Copyright © 2008 Elsevier Inc. All rights reserved

No part of this publication may be reproduced, stored in a retrieval system or transmitted in any form or by any means electronic, mechanical, photocopying, recording or otherwise without the prior written permission of the publisher

Permissions may be sought directly from Elsevier's Science & Technology Rights Department in Oxford, UK: phone (+44) (0) 1865 843830; fax (+44) (0) 1865 853333; email: permissions@elsevier.com. Alternatively you can submit your request online by visiting the Elsevier web site at http://elsevier.com/locate/permissions, and selecting *Obtaining permission to use Elsevier material*

Notice
No responsibility is assumed by the publisher for any injury and/or damage to persons or property as a matter of products liability, negligence or otherwise, or from any use or operation of any methods, products, instructions or ideas contained in the material herein. Because of rapid advances in the medical sciences, in particular, independent verification of diagnoses and drug dosages should be made

ISBN: 978-0-12-374326-8
ISSN: 0065-2776

For information on all Academic Press publications
visit our website at elsevierdirect.com

Printed and bound in USA
08 09 10 11 10 9 8 7 6 5 4 3 2 1

**Working together to grow
libraries in developing countries**

www.elsevier.com | www.bookaid.org | www.sabre.org

ELSEVIER   BOOK AID International   Sabre Foundation

CONTENTS

Contributors ix

1. **Autoimmune Diabetes Mellitus—Much Progress, but Many Challenges** 1

    Hugh O. McDevitt and Emil R. Unanue

    1. Models of Type 1 DM 2
    2. The Histocompatibility Molecules of the Autoimmune Diabetic 3
    3. The Diabetic Antigens and the Diabetogenic T Cells 3
    4. The Genetics of Type 1 Diabetes 6
    5. Development of New Therapies in the NOD Model Which Show Promise in Studies in Patients 7
    6. Effects of Administration of Glutamic Acid Decarboxylase 65 (GAD 65) to Patients with Recent-Onset T1D 8
    7. The Prospect for Trials to Prevent T1DM in Susceptible Human Populations 9
    References 10

2. **CD3 Antibodies as Unique Tools to Restore Self-Tolerance in Established Autoimmunity: Their Mode of Action and Clinical Application in Type 1 Diabetes** 13

    Sylvaine You, Sophie Candon, Chantal Kuhn, Jean-François Bach, and Lucienne Chatenoud

    1. Introduction 14
    2. CD3 Antibodies as Promising Tolerance-Promoting Tools: The Proof of Concept in T1D 16
    3. The Clinical Results in T1D: More than 25 Years of Effort 18
    4. Why is The Effect of CD3-Antibody Based Therapy Antigen Specific and Long Lasting? 21
    5. Building on the Present Clinical Experience to Improve CD3 Antibody Therapy in T1D 26
    6. Conclusions 30
    References 30

3. GAD65 Autoimmunity—Clinical Studies   39

Raivo Uibo and Åke Lernmark

1. Introduction	40
2. GAD65 Autoimmunity in T1D	48
3. Pathogenesis of GAD65 Autoimmunity	61
4. Preclinical GAD65 Studies	64
5. Phase I GAD65 Clinical Studies	64
6. Phase II GAD65 Clinical Studies	65
7. Ongoing and Future Clinical Studies	67
8. Concluding Remarks	68
Acknowledgments	68
References	68

4. CD8+ T Cells in Type 1 Diabetes   79

Sue Tsai, Afshin Shameli, and Pere Santamaria

1. Introduction	80
2. MHC Class I and T1D	81
3. Autoreactive CD8+ T Cells in T1D	83
4. Antigens for Diabetogenic CD8+ T Cells in Humans	86
5. The Relative Contribution of the β-Cell-Specific CD8+ Response to T1D	95
6. Development and Activation of Diabetogenic CD8+ T Cells	96
7. Recruitment of Diabetogenic CD8+ T Cells to Islets	104
8. Mechanisms of β-Cell Cytotoxicity in T1D	105
9. Induction of Immunologic Tolerance in Diabetogenic CD8+ T Cells	108
10. Concluding Remarks	111
Acknowledgments	112
References	112

5. Dysregulation of T Cell Peripheral Tolerance in Type 1 Diabetes   125

R. Tisch and B. Wang

1. Introduction	126
2. The Autoimmune Process of T1D	126
3. Dysregulation of Central T Cell Tolerance in T1D	128
4. Dysregulation of Peripheral T Cell Tolerance in T1D	130
5. Summary	140
Acknowledgments	141
References	141

6. Gene–Gene Interactions in the NOD Mouse Model of Type 1 Diabetes   151

William M. Ridgway, Laurence B. Peterson, John A. Todd,
Dan B. Rainbow, Barry Healy, Oliver S. Burren, and Linda S. Wicker

1. Type 1 Diabetes is a Multigenic Disease	152
2. Insights into a Multigenic Disease: The NOD Mouse	153
3. Defining the Function of Alleles Altering the Frequency of T1D	161
4. Functional Studies with NOD Congenic Mice: The Combination of Protective Alleles at *Idd3* and *Idd5* Prevent Diabetes and Insulitis	162
5. Review of *Idd3*	163
6. Review of *Idd5*	165
7. Genetic Complexity of *Idd5*: Genetic Masking and the Discovery of *Idd5*.3 and *Idd5*.4	167
8. Combining Protective Alleles at Il2/*Idd3* with those from the *Idd5* Subregions	168
9. Conclusion	169
Acknowledgements	170
References	170

Index   177
Content of Recent Volumes   181
See Color Plate Section in the back of this book

CONTRIBUTORS

Numbers in parentheses indicate the pages on which the authors' contributions begin.

Jean-François Bach
INSERM, Unité 580 and Université Paris Descartes, 75015 Paris, France (13)

Oliver S. Burren
Juvenile Diabetes Research Foundation/Wellcome Trust Diabetes and Inflammation Laboratory, Cambridge Institute for Medical Research, Wellcome Trust/MRC Building, Addenbrooke's Hospital, Cambridge CB2 0XY, United Kingdom (151)

Sophie Candon
INSERM, Unité 580 and Université Paris Descartes, 75015 Paris, France (13)

Lucienne Chatenoud
INSERM, Unité 580 and Université Paris Descartes, 75015 Paris, France (13)

Barry Healy
Juvenile Diabetes Research Foundation/Wellcome Trust Diabetes and Inflammation Laboratory, Cambridge Institute for Medical Research, Wellcome Trust/MRC Building, Addenbrooke's Hospital, Cambridge CB2 0XY, United Kingdom (151)

Chantal Kuhn
INSERM, Unité 580 and Université Paris Descartes, 75015 Paris, France (13)

Åke Lernmark
Lund University/CRC, Department of Clinical Sciences, University Hospital MAS, 20502 Malmö, Sweden (39)

Hugh O. McDevitt
Department of Microbiology and Immunology, School of Medicine, Stanford University, Palo Alto, California, USA (1)

Laurence B. Peterson
Merck Research Laboratories, Rahway, New Jersey, USA (151)

Dan B. Rainbow
Juvenile Diabetes Research Foundation/Wellcome Trust Diabetes and Inflammation Laboratory, Cambridge Institute for Medical Research, Wellcome Trust/MRC Building, Addenbrooke's Hospital, Cambridge CB2 0XY, United Kingdom (151)

William M. Ridgway
University of Pittsburgh School of Medicine, 725 SBST, Pittsburgh, Pennsylvania, USA (151)

Pere Santamaria
Julia McFarlane Diabetes Research Centre (JMDRC), Department of Microbiology and Infectious Diseases, and Calvin, Phoebe and Joan Snyder Institute of Infection, Immunity and Inflammation, Faculty of Medicine, University of Calgary, Calgary, Alberta, Canada (79)

Afshin Shameli
Julia McFarlane Diabetes Research Centre (JMDRC), Department of Microbiology and Infectious Diseases, and Calvin, Phoebe and Joan Snyder Institute of Infection, Immunity and Inflammation, Faculty of Medicine, University of Calgary, Calgary, Alberta, Canada (79)

R. Tisch
Department of Microbiology and Immunology, University of North Carolina at Chapel Hill, Chapel Hill, North Carolina, USA (125)

John A. Todd
Juvenile Diabetes Research Foundation/Wellcome Trust Diabetes and Inflammation Laboratory, Cambridge Institute for Medical Research, Wellcome Trust/MRC Building, Addenbrooke's Hospital, Cambridge CB2 0XY, United Kingdom (151)

Sue Tsai
Julia McFarlane Diabetes Research Centre (JMDRC), Department of Microbiology and Infectious Diseases, and Calvin, Phoebe and Joan Snyder Institute of Infection, Immunity and Inflammation, Faculty of Medicine, University of Calgary, Calgary, Alberta, Canada (79)

Raivo Uibo
Department of Immunology, IGMP, Centre of Molecular and Clinical Medicine, University of Tartu, Tartu, Estonia (39)

Emil R. Unanue
Department of Pathology and Immunology, Washington University School of Medicine, St. Louis, Missouri, USA (1)

B. Wang
Department of Microbiology and Immunology, University of North Carolina at Chapel Hill, Chapel Hill, North Carolina, USA (125)

Linda S. Wicker
Juvenile Diabetes Research Foundation/Wellcome Trust Diabetes and Inflammation Laboratory, Cambridge Institute for Medical Research, Wellcome Trust/MRC Building, Addenbrooke's Hospital, Cambridge CB2 0XY, United Kingdom (151)

Sylvaine You
INSERM, Unité 580 and Université Paris Descartes, 75015 Paris, France (13)

CHAPTER 1

Autoimmune Diabetes Mellitus—Much Progress, but Many Challenges

Hugh O. McDevitt* and **Emil R. Unanue**[†]

Contents		
	1. Models of Type 1 DM	2
	2. The Histocompatibility Molecules of the Autoimmune Diabetic	3
	3. The Diabetic Antigens and the Diabetogenic T Cells	3
	4. The Genetics of Type 1 Diabetes	6
	5. Development of New Therapies in the NOD Model Which Show Promise in Studies in Patients	7
	6. Effects of Administration of Glutamic Acid Decarboxylase 65 (GAD 65) to Patients with Recent-Onset T1D	8
	7. The Prospect for Trials to Prevent T1DM in Susceptible Human Populations	9
	References	10

The past 10 years have seen major advances in our understanding of several key areas of research on the pathogenesis of Type 1 diabetes mellitus (T1DM) in both man and the principal animal model, the non-obese diabetic (NOD) mouse (Hattori *et al.*, 1986; Kikutani and Makino, 1992). As will be discussed in the following sections, particularly in the genetic studies led by Linda Wicker, extensive insights have been gleaned from the NOD model. Many of these have subsequently been found in patients with both prediabetes and recent-onset diabetes. Further, new

---

* Department of Microbiology and Immunology, School of Medicine, Stanford University, Palo Alto, California
[†] Department of Pathology and Immunology, Washington University School of Medicine, St. Louis, Missouri

therapeutic approaches have been developed in the NOD model, and some of these have been tested in patients with recent-onset T1DM. This has created a situation in which attempts to induce long-term prevention can be contemplated in much greater detail than heretofore.

T1DM or autoimmune diabetes is a complex, chronic autoimmune disease with a serious clinical impact (Bach, 1994; Castano and Eisenbarth, 1990; Tisch and McDevitt, 1966). It is estimated that approximately 1–1.5 million patients have T1DM in the United States. The incidence is increasing, as shown from studies in several counties in Massachusetts in the mid-1940s and again in the 1980s. During that time there has been a rough doubling of the incidence of diabetes in this relatively stable population.

1. MODELS OF TYPE 1 DM

Extensive studies, both on the human disease and on the various experimental models, have taken place in the last several years. In this issue, five contributors to this book on autoimmune T1DM have written chapters that cover their work and those of others. A major emphasis of some of them concerns the basic features of the autoimmune process, as examined in the spontaneous disease in the NOD mouse. Herewith are brief comments on some of our perspectives on this fascinating and challenging autoimmune process.

Research in autoimmune diabetes was encouraged by a number of animal models that reproduced the human disease. The earliest model that gave an indication of an autodestructive process involved the use of the drug streptozotocin. Streptozotocin, an alkylating agent with a structure akin to glucose, when given to mice in a large dose (about 200 mg per kilogram of body weight), rapidly destroyed β cells in 24–48 h. However, a much smaller dose, about one-fifth of the large dose, resulted in a slow development of diabetes with the presence of an early inflammatory response. From a number of studies, Rossini *et al.* concluded that a likely autoimmune reactivity was responsible for the perpetuation of the inflammation (Rossini *et al.*, 1977). Whether streptozotocin induces T cells that cause diabetes can be disputed, but regardless, their findings were the first that experimentally related the inflammation of islets to diabetes. Notable advances in the experimental field came as a result of the availability of two genetic strains that developed spontaneous diabetes: one, the rat BioBreeder or BB strain; the other, the NOD mouse. The BB rat has given much information on the pathology and regulatory mechanisms (Mordes *et al.*, 2007). However, at this time, research in the NOD mouse has superseded that in the BB rat, in great part because of the genetic similarities between the mouse and human diabetes, plus the ease of manipulation of the mouse strain.

The use of the NOD mouse has been extensive (Tisch and McDevitt, 1996). It is one of the few autoimmune diseases where the process develops spontaneously without the need of immunization, as is required in most other autoimmune disease models; for example, the experimental models of multiple sclerosis or of autoimmune myocarditis. Depending on the housing conditions, from 60% to 90% of female NOD mice will develop diabetes, starting about the 10th–12th week of life.

2. THE HISTOCOMPATIBILITY MOLECULES OF THE AUTOIMMUNE DIABETIC

The class II major histocompatibility (MHC) molecule of the NOD is I-A^{g7} (Acha-Orbea and McDevitt, 1987). In T1DM patients, a genetic linkage with the HLA-DQ8 molecule was found in many patients, a finding first made by one of us (Todd et al., 1987). Both molecules exhibit a remarkable similarity: the presence of a nonaspartic residue at the 57 residue of the β chain.

Both the human DQ8 molecule and the murine I-A^{g7} have been crystallized, and binding interactions with peptides have been studied (Corper et al., 2000; Latek et al., 2000; Lee et al., 2001). By utilizing mass spectrometry approaches, the chemical features of peptides bound to these class II MHC molecules of antigen presenting cells (APC) have shown a marked preference for acidic residues at the carboxy-end of the peptides at the P9 position (reviewed in Suri et al., 2008). The acidic residues interact with the unpaired arginine at the α chain 76 residue. Up to 90% of peptides eluted from either DQ8 or I-A^{g7} molecules will contain one or more acidic residues at the P9 position. The other two residues encountered among the peptides have either alanines or glycines at P9. These findings set a base in searching and examining β cell derived peptides linked to these two allelic forms of class II MHC molecules.

3. THE DIABETIC ANTIGENS AND THE DIABETOGENIC T CELLS

Agreeing that CD4 T cells are central to the autoimmune process, searches are underway by several laboratories for the identification of antigens presented by the diabetes susceptible class II MHC molecules, and which induce CD4 diabetogenic T cells. As expected, most of this work has centered on the NOD mouse where isolation of the T cells in draining nodes and infiltrated islets is possible, as is their testing by transferring the T cells to NOD recipients (Haskins et al., 1988). Recent reviews have discussed this issue (Haskins, 2005; Lieberman and DiLorenzo, 2003). One finding that is striking is the diversity of the autoimmune repertoire of the NOD mouse. At least 15 distinct β cell derived peptides, and the corresponding CD4 T cells, have been identified among various laboratories. The searches have followed different approaches: testing T cell lines or clones on proteins known to be part of the β cell secretory granules and (or) proteins first identified by autoantibodies. This approach was highly successful in the identification of T cells directed to insulin (Wegmann et al., 1994) and GAD (Tisch and McDevitt, 1993), the two proteins most extensively examined.

The presentation of β cell antigens is a complex issue in which β cells do not express class II MHC molecules. Presentation takes place by transfer of β cell antigens to the APC. Islets of Langerhans contain dendritic cells (DCs), and these bear peptide-MHC complexes derived from β cell granules (Calderon et al., 2008). Evidence is strong that the lymph nodes draining the pancreas are crucial to the selection and activation of diabetogenic T cells (Gagnerault et al., 2002). How the β cell antigen presentation takes place is an important issue to investigate, i.e., whether by migration of islet DC to the lymph nodes or by drainage of β cell products directly to the nodes to be taken up by their DC. Although DCs are viewed as the essential APC, an important role for B cells as APC has been well documented (Serreze et al., 1998).

There are several important issues to examine concerning the repertoire of T cells.

First, despite finding many T cells, there are still a considerable number of diabetogenic T cells isolated from diabetic or prediabetic mice whose antigen specificity is not known. Indeed, among these are many of the "BDC" clones from the series isolated by Katie Haskins (Haskins et al., 1988). It is of interest that a substantial number reacted with a mimotope peptide that activated the BDC2.5 T cell, isolated by the Haskins laboratory and then made into a transgenic T cell receptor by Diane Mathis and Christophe Benoist (Katz et al., 1993). Identifying the specificity from the many β cell specific proteins constitutes a problem. In our hands (ERU) testing T cells isolated from diabetic mice against peptides from β cell granules has not been effective, perhaps because of the many proteins and peptides that are involved.

Second, testing the diabetogenicity of the T cells has been a problem. Many T cells transfer a mild insulitis, or none at all, raising the question whether this poor response is a question of antigen specificity or of the features of a particular T cell. One can speculate that there are autoantigens that drive the initial stage of the disease, and that once inflammation sets in, "epitope spreading" takes place. A dominant autoantigen may be one that is heavily expressed in β cells. Haskins has analyzed cogently her experience (Haskins, 2005): the problem remains in interpreting a negative result. Despite these problems, some T cells are reproducibly diabetogenic; that is, they can transfer diabetes.

The most notable example has been the transfer of diabetes by anti-insulin T cells, first reported by Wegmann and coworkers (Daniel et al., 1995; Wegmann et al., 1994) and further examined by others, particularly by Eisenbarth's research group (e.g., Moriyama et al., 2003; Zhang et al., 2008). Insulin reactive T cells are identified early in pancreatic lymph nodes of NOD mice. In humans, the antibody response to insulin is an excellent marker of the disease, and T cells to insulin have also been identified. That insulin appears to be a major autoantigen is strongly supported by experiments carried out in Harrison and von Boehmer's laboratory (French et al., 1997; Jaeckel et al., 2007). They tackled the role of a diabetic antigen by presenting it in the thymus gland, a manipulation that was permitted by expressing the antigen in class II gene constructs: mice that expressed insulin by thymic APC had a reduced or complete absence of disease.

Particularly pointed are the negative results obtained; that is, no effect on the autoimmune process when other autoantigens are expressed in thymic cells. Such has been the experience when GAD was expressed. The GAD proteins trigger autoantibodies, a finding cogently analyzed in Chapter 3 by Ake Lernmark, this volume.

Third, the issue of NOD autoimmunity extends beyond diabetes. Other autoimmune traits have been identified in NOD, including sialitis, myocarditis, and prostatitis, for example. These findings point to the NOD mouse as having some defect in tolerance to self-proteins, favoring autoimmune responses. Roland Tisch and Bo Wang have made a comprehensive review of this important feature of the NOD mouse, summarizing the many observations that have now been reported on the possible cellular basis, that is, regulatory T cells, antigen presentation problems, and dendritic cells defects. These studies revealed a central thymic problem in negative selection in the NOD mouse (Kishimoto and Sprent, 2001; Liston et al., 2004). We argue that the particular biochemical features of the class II molecule I-A^{g7}, combined with the lack of expression of I-E (the murine homolog of HLA DR) are the factors that result in the propensity for a more florid autoimmune response in NOD mice (Suri et al., 2008). In an early study, T cells from NOD mice were shown to develop a strong autoimmune response to self antigens expressed by their own APC (Kanagawa et al., 1998; Ridgway et al., 1998). For example, the autologous MLR response was strong in NOD mice (which express I-A^{g7}), but weak in NOD expressing I-A^b; in contrast, B6 mice expressing I-A^{g7} exhibited a strong autologous MLR. What could be peculiar about I-A^{g7} that favors such prominent autoreactivity is not clear. We have discussed that I-A^{g7} tends to bind peptides weakly and is an unstable molecule, traits that could favor a defect in negative selection. It is of interest to note that many of the β cell autoantigens that trigger a T cell response bind to I-A^{g7} weakly, have a fast dissociation rate, and yet are diabetogenic (reviewed in Suri *et al.*, 2008).

A notable feature of NOD diabetogenesis is the involvement of other cells in the process, either as effector cells or as cells that influence or regulate the process. In this issue, Pere Santamaria makes a cogent analysis of the importance of CD8 T cells in the diabetogenic process, covering all the various aspects of their biology and specificity. That CD8 T cells contributed significantly to the diabetic process was first brought out in early studies in which spleen cells from diabetic NOD were transferred into nondiabetic mice. CD8 T cells not only contributed, but frequently were required, together with CD4 T cells. Notable insights have been gathered on the involvement of CD8 T cells, particularly with the identification of the IGRP antigen as a target of them, and the development of CD8 TCR transgenic mice.

4. THE GENETICS OF TYPE 1 DIABETES

Genome wide association studies combined with single nucleotide polymorphisms (SNP) studies identified a number of genes in the NOD mouse that predispose to Type 1 diabetes. The recent studies described by Ridgway *et al.* (Chapter 6, this volume) have revealed striking findings. Initial studies in the NOD mouse and in human diabetic and control populations identified a large number of genes that either predispose to the development of Type 1 diabetes, or in cases such as HLA DR2, are protective against the development of T1DM, even in patients with other susceptibility genes. Current estimates in the NOD mouse indicate that ~60% of the inheritance of susceptibility is due to genes in the MHC (Chapter 6 by Ridgway *et al.*, this volume). Subsequent to these initial studies, Todd's laboratory (Todd *et al.*, 1991), and later Wicker and Todd and their associates (Wicker *et al.*, 2005; Chapter 6 by Ridgway *et al.*, this volume), identified some 20 chromosomal regions in the NOD mouse that predispose to the development of T1DM. In order to analyze these genes and their effects in detail, Wicker *et al.* created a series of congenic strains that carry Idd3, Idd5, and a number of other diabetes susceptibility regions from the NOD mouse genome on a standard genetic background. This permitted the analysis of the effects of these genes, the incidence of development of T1DM by various combinations, and an understanding of how these genes interact to mediate susceptibility to T1DM.

These studies are described in detail in Chapter 6 by Ridgway *et al.*, this volume. It now appears clear from identification of a number of genetic variants which are involved in the autoimmune pathogenesis in the NOD mouse, and from linkage studies, that this approach permits insights into the molecular consequences of different alleles mediating protection or susceptibility. In parallel, the identification of candidate genes from human genome-wide association studies (in addition to those Type 1 diabetes genes already discovered in human) underscores a common pathogenesis for human and mouse T1DM. This emphasizes the relevance of the NOD strain for modeling the diabetic process in human. Indeed, the close similarity/identity of the relatively small number of genes that have been identified as major players in diabetes susceptibility in the NOD mouse and in human indicates that the pathogenesis of this disease in these two species is very similar. Thus, findings initially made in the NOD model can be extrapolated to human. Further, the analysis of these genes and the identification of the precise gene products and their involvement in pathways leading to inflammation and β islet cell destruction open the way to identifying new targets that could be evaluated as therapeutic targets for prevention of T1DM.

5. DEVELOPMENT OF NEW THERAPIES IN THE NOD MODEL WHICH SHOW PROMISE IN STUDIES IN PATIENTS

Chatenoud *et al.* found a striking effect of the administration of antiCD3 to NOD mice at the first onset of glycosuria (Chatenoud *et al.*, 1994, 1997; reviewed in Chapter 2 by You *et al.*, this volume). These authors studied the effects of this treatment with a monoclonal antibody to CD3, a T-cell protein found in all T cells. They found that administration of the monoclonal antibody just after birth, or in the weeks preceding the actual onset of diabetes, had little effect. However, administration just at the first onset of glycosuria resulted in a decrease in glycosuria, return of the blood sugar to normal levels, and complete prevention of diabetes for periods up to 1 year. Extensive studies of the mechanism of this effect are described in the article by Chatenoud *et al.*, and appear to be due to the induction of regulatory T cells that change the nature of the islet protein specific T-cell repertoire to one of down regulation, rather than one of inflammation (Chatenoud, 2005; Chatenoud and Bluestone, 2007; Chapter 2 by You *et al.*, this volume).

These studies have been extended in man with the administration of antibodies to human CD3 engineered to lack binding to the Fc receptor, and also lacking the complement-binding site, resulting in an antibody which does not induce the tremendous cytokine response that is seen with unmodified monoclonal antiCD3. These studies were carried out in recent-onset T1DM patients (Herold *et al.*, 2002, 2005; Keymeulen *et al.*, 2005).

Administration of this antibody within a few weeks of the initial onset of glycosuria has resulted in marked preservation of ability to produce C-peptide and a decrease in the insulin requirement compared to controls who did not receive this monoclonal antibody. Follow-up studies have shown that this preservation of C-peptide levels, indicating preservation of a residual small amount of β cell mass, persists for as long as 5 years after the therapeutic treatment described above. Thus, there is the likelihood that this type of therapy could be effective in preventing Type 1 diabetes if administered at the proper time and in a proper dose. Unfortunately, the studies were complicated by the occurrence of a syndrome resembling infectious mononucleosis, with a febrile illness lasting several weeks, with moderate morbidity.

6. EFFECTS OF ADMINISTRATION OF GLUTAMIC ACID DECARBOXYLASE 65 (GAD 65) TO PATIENTS WITH RECENT-ONSET T1D

Uibo and Lernmark (Chapter 3, this volume) review the autoimmune response to GAD 65 found in T1DM patients and NOD mice. Early studies utilizing recombinant-produced, purified GAD 65 in NOD mice with well-established insulitis, but without signs of glycosuria, revealed that it was possible to reduce the incidence of T1DM in female NOD mice treated at 12 weeks from 80% to 16–18%. This prevention lasted throughout the experiment, which extended to 36 weeks. As with antiCD3, this treatment was extended to patients with recent onset diabetes by a Swedish firm, Dyamid, utilizing purified GAD 65 isolated from bioengineered *Escherichia coli*. A protective effect lasted for at least 18 months and is now being further tested for its duration. Of note, the patients in the initial study who showed preservation of C-peptide levels and a decrease in insulin requirement were the patients with the shortest time lapse between onset of glycosuria and administration of GAD 65. Patients with recent-onset diabetes occurring more than 6 months prior to treatment with GAD 65 showed no therapeutic effect. It is also of interest that lower doses and higher doses of GAD 65 (up to 100 μg) were not as effective as the 20 g dose which was the one used in the trial. Thus, it appears that this particular dose of GAD 65 given to patients with recent onset T1DM confers the ability to maintain C-peptide levels, apparently by inducing regulatory T cells that prevent the progression of the diabetic process. This results in a population of patients with long-term lower requirements for insulin. This therapy has not yet been tested in prevention. It should also be noted that there were essentially no adverse effects of this method of treatment.

7. THE PROSPECT FOR TRIALS TO PREVENT T1DM IN SUSCEPTIBLE HUMAN POPULATIONS

The studies described in this volume show that our understanding of the pathogenesis by which both CD4 and CD8 T cells collaborate to destroy β islet cells is sufficiently far advanced that it should be possible to contemplate methods of preventing the disease. It is now possible to identify patients at risk for subsequent development of diabetes by determining the presence or absence of significant titers of antibodies to insulin, GAD, and IA-2. In individuals with a known susceptible HLA genotype, presence of any two of these three autoantibodies in significant titers predicts progression to Type 1 diabetes in 80–90% of patients within 5 years (reviewed in Eisenbarth *et al.*, 2002; and Chapter 3 by Uibo and Lernmark, this volume).

Methods are also available for assessing actual β cell destruction by estimating β cell mass. This can be done by careful study of glucose metabolism and insulin reserve followed longitudinally in patients who have the relevant autoantibodies and a susceptible HLA genotype. These parameters will predict patients that are high and low risk for development of T1DM. Once patients with a susceptible HLA genotype, significant autoantibody titers to any two of the three major islet cell autoantigens, and clear evidence of a diminution in insulin reserve have been documented, these patients will be the ideal group for an attempt to prevent further progression of the process. As preventive therapies are developed further, such information will be vital: one needs to be sure that such therapies, which are not without risks, will be given to patients with a high probability of developing the disease. This is because the concordance rate for Type 1 diabetes in identical twins is not 100%. It is ~70% in identical twins expressing both HLA DR3 and DR4. In other HLA genotypes, the concordance rate is lower, thus raising the possibility of treating individuals who are not at risk, unless significant β cell loss is documented.

There are a number of considerations that need to be discussed in any clinical endeavor. Perhaps the most important is the necessity for identifying treatments in which the morbidity is extremely low and adverse events are nonexistent. The use of monoclonal antibodies for long-term therapies will need to be critically considered, as was discussed here in the review by Chatenoud's group (Chapter 2 by You *et al.*, this volume). For example, the problem of reactivation of EBV infection following antiCD3 therapy raises the question of whether acute or long-term adverse effects might follow this type of therapy. Therapies using autoantigens known to be involved in the autoimmune process will continue to be examined; some have failed, others have given some encouraging results. These will become more prevalent as our knowledge of diabetic autoantigens increases, their role at different stages is better understood, as well as their involvement in regulatory T cell pathways. These are matters which must be discussed and argued extensively within the community of investigators involved in studies of Type 1 diabetes; the relevant personnel in the Food & Drug Administration; and the relevant human ethics committees. Nonetheless, the studies presented in this volume raise the possibility that the time to consider studies on prevention of T1D may have arrived.

REFERENCES

Acha-Orbea, H., and McDevitt, H. O. (1987). The first external domain of the nonobese diabetic mouse class II I-A beta chain is unique. *Proc. Natl. Acad. Sci. USA* **84,** 2435–2439.

Bach, J. F. (1994). Insulin-dependent diabetes mellitus as an autoimmune disease. *Endocr. Rev.* **15,** 516–542.

Calderon, B., Suri, A., Miller, M. J., and Unanue, E. R. (2008). Dendritic cells in islets of Langerhans constitutively present beta cell derived peptides bound to their class II MHC molecules. *Proc. Natl. Acad. Sci. USA* **105,** 6121–6126.

Castano, L., and Eisenbarth, G. S. (1990). Type-I diabetes: A chronic autoimmune disease of human, mouse, and rat. *Annu. Rev. Immunol.* **8,** 647–649.

Chao, C. C., and McDevitt, H. O. (1997). Identification of immunogenic epitopes of GAD 65 presented by Ag7 in nonobese diabetic mice. *Immunogenetics* **46,** 29–34.

Chatenoud, L. (2005). CD3-specific antibodies restore self-tolerance: Mechanisms and clinical applications. *Curr. Opin. Immunol.* **17,** 632–637.

Chatenoud, L., and Bluestone, J. A. (2007). CD3-specific antibodies: A portal to the treatment of autoimmunity. *Nat. Rev. Immunol.* **7,** 622–632.

Chatenoud, L., Thervet, E., Primo, J., and Bach, J. F. (1994). Anti-CD3 antibody induces long-term remission of overt autoimmunity in nonobese diabetic mice. *Proc. Natl. Acad. Sci. USA* **91,** 123–127.

Chatenoud, L., Primo, J., and Bach, J. F. (1997). CD3 antibody-induced dominant self tolerance in overtly diabetic NOD mice. *J. Immunol.* **158,** 2947–2954.

Corper, A. L., Stratmann, T., Apostolopoulos, V., Scott, C. A., Garcia, K. C., Kang, A. S., Wilson, I. A., and Teyton, L. (2000). A structural framework for deciphering the link between I-Ag7 and autoimmune diabetes. *Science* **288,** 505–511.

Daniel, D., Gill, R. G., Schloot, N., and Wegmann, D. (1995). Epitope specificity, cytokine production profile and diabetogenic activity of insulin-specific T cell clones isolated from NOD mice. *Eur. J. Immunol.* **25,** 1056–1062.

Eisenbarth, G. S., Moriyama, H., Robles, D. T., Liu, E., Yu, L., Babu, S., Redondo, M. J., Gottlieb, P., Wegmann, D., and Rewers, M. (2002). Insulin autoimmunity: Prediction/precipitation/prevention type 1A diabetes. *Autoimmun. Rev.* **1,** 139–145.

French, M. B., Allison, J., Cram, D. S., Thomas, H. E., Dempsey-Collier, M., Silva, A., Georgiou, H. M., Kay, T. W., Harrison, L. C., and Lew, A. M. (1997). Transgenic expression of mouse proinsulin II prevents diabetes in nonobese diabetic mice. *Diabetes* **46,** 34–39.

Gagnerault, M. C., Luan, J. J., Lotton, C., and Lepault, F. (2002). Pancreatic lymph nodes are required for priming of β cell reactive T cells in NOD mice. *J. Exp. Med.* **196,** 369–377.

Godkin, A., Friede, T., Davenport, M., Stevanovic, S., Willis, A., Jewell, D., Hill, A., and Rammensee, H. G. (1997). Use of eluted peptide sequence data to identify the binding characteristics of peptides to the insulin-dependent diabetes susceptibility allele HLA-DQ8 (DQ 3.2). *Int. Immunol.* **9,** 905–911.

Haskins, K. (2005). Pathogenic T-cell clones in autoimmune diabetes: More lessons from the NOD mouse. *Adv. Immunol.* **87,** 123–162.

Haskins, K., Portas, M., Bradley, B., Wegmann, D., and Lafferty, K. (1988). T-lymphocyte clone specific for pancreatic islet antigen. *Diabetes* **37,** 1444–1448.

Hattori, M., Buse, J. B., Jackson, R. A., Glimcher, L., Dorf, M. E., Minami, M., Makino, S., Moriwaki, K., Kuzuya, H., Imura, H., *et al.* (1986). The NOD mouse: Recessive diabetogenic gene in the major histocompatibility complex. *Science* **231,** 733–735.

Herold, K. C., Hagopian, W., Auger, J. A., Poumian-Ruiz, E., Taylor, L., Donaldson, D., Gitelman, S. F., Harlan, D. M., Xu, D., Zivin, R. A., and Bluestone, J. A. (2002). Anti-CD3 monoclonal antibody in new onset type 1 diabetes mellitus. *N. Engl. J. Med.* **346,** 1692–1698.

Herold, K. C., Gitelman, S. E., Masharani, U., Hagopian, W., Bisikirska, B., Donaldson, D., Rother, K., Diamond, B., Harlan, D. M., and Bluestone, J. A. (2005). A single course of anti-CD3 monoclonal antibody hOKT3gamma1 (Ala-Ala) results in improvement in C-peptide responses and clinical parameters for at least 2 years after onset of type 1 diabetes. *Diabetes* **54,** 1763–1769.

Jaeckel, E., Lipes, M. A., and von Boehmer, H. (2004). Recessive tolerance to preproinsulin 2 reduces but does not abolish type 1 diabetes. *Nat. Immunol.* **5,** 1028–1035.

Kanagawa, O., Martin, S. M., Vaupel, B. A., Carrasco-Marin, E., and Unanue, E. R. (1998). Autoreactivity of T cells from nonobese diabetic mice: An I-A^{g7}- dependent reaction. *Proc. Natl. Acad. Sci. USA* **95,** 1721–1724.

Katz, J. D., Wang, B., Haskins, K., Benoist, C., and Mathis, D. (1993). Following a diabetogenic T cell from genesis through pathogenesis. *Cell* **74,** 1089–1100.

Keymeulen, B., Vandemeulebroucke, E., Ziegler, A. G., Mathieu, C., Kaufman, L., Hale, G., Gorus, F., Goldman, M., Walter, M., Candon, S., Schandene, L., Crenier, L., *et al.* (2005). Insulin needs after CD3-antibody therapy in new-onset type 1 diabetes.

Kikutani, H., and Makino, S. (1992). The murine autoimmune diabetes model: NOD and related strains. *Adv. Immunol.* **51,** 285–322.

Kishimoto, H., and Sprent, J. (2001). A defect in central tolerance in NOD mice. *Nat. Immunol.* **2,** 1025–1031.

Latek, R. R., Suri, A., Petzold, S. J., Nelson, C. A., Kanagawa, O., Unanue, E. R., and Fremont, D. H. (2000). Structural basis of peptide binding and presentation by the type I diabetes-associated MHC class II molecule of NOD mice. *Immunity* **12,** 699–710.

Lee, K. H., Wucherpfennig, K. W., and Wiley, D. C. (2001). Structure of a human insulin peptide-HLA-DQ8 complex and susceptibility to type 1 diabetes. *Nat. Immunol.* **2,** 501–507.

Lieberman, S. M., and DiLorenzo, T. P. (2003). A comprehensive guide to antibody and T-cell responses in type 1 diabetes. *Tissue Antigens* **62,** 359–377.

Liston, A., Lesage, S., Gray, D. H., O'Reilly, L. A., Strasser, A., Fahrer, A. M., Boyd, R. L., Wilson, J., Baxter, A. G., Gallo, E. M., Crabtree, G. R., Peng, K., *et al.* (2004). Generalized resistance to thymic deletion in the NOD mouse; a polygenic trait characterized by defective induction of Bim. *Immunity* **21,** 817–830.

Mordes, J. P., Poussier, P., Rossini, A. A., Blankenhorn, E. P., and Greiner, D. L. (2007). Rat models of type 1 diabetes: Genetics, environment, and autoimmunity. *In* "*Animal Models of Diabetes. Frontiers in Research*" (Shafrir Eleazar, Ed.), 2nd edition., pp. 1–39. CRC Press, Boca Raton.

Moriyama, H., Abiru, N., Paronen, J., Sikora, K., Liu, E., Miao, D., Devendra, D., Beilke, J., Gianani, R., Gill, R. G., and Eisenbarth, G. S. (2003). Evidence for a primary islet autoantigen (preproinsulin 1) for insulitis and diabetes in the nonobese diabetic mouse. *Proc. Natl. Acad. Sci. USA* **100**, 10376–10381.

Ridgway, W. M., Ito, H., Fasso, M., Yu, C., and Fathman, C. G. (1998). Analysis of the role of variation of major histocompatibility complex class II expression on nonobese diabetic (NOD) peripheral T cell response. *J. Exp. Med.* **188**, 2267–2275.

Rossini, A. A., Like, A. A., Chick, W. L., Appel, M. C., and Cahill, G. F., Jr. (1977). Studies of streptozotocin-induced insulitis and diabetes. *Proc. Natl. Acad. Sci. USA* **74**, 2485–2489.

Serreze, D. V., Fleming, S. A., Chapman, H. D., Richard, S. D., Leiter, E. H., and Tisch, R. M. (1998). B lymphocytes are critical antigen-presenting cells for the initiation of T cell-mediated autoimmune diabetes in nonobese diabetic mice. *J. Immunol.* **161**, 3912–3918.

Suri, A., Levisetti, M. G., and Unanue, E. R. (2008). Do the peptide-binding properties of diabetogenic class II molecules explain autoreactivity? *Curr. Opin. Immunol* **20**, 105–110.

Tisch, R., and McDevitt, H. (1996). Insulin-dependent diabetes mellitus. *Cell* **85**, 291–297.

Todd, J. A., Bell, J. I., and McDevitt, H. O. (1987). HLA-DQ beta gene contributes to susceptibility and resistance to insulin-dependent diabetes mellitus. *Nature* **329**, 599–604.

Todd, J. A., Aitman, T. J., Cornall, R. J., Ghosh, S., Hall, J. R., Hearne, C. M., Knight, A. M., Love, J. M., McAleer, M. A., Prins, J. B., *et al.* (1991). Genetic analysis of autoimmune type 1 diabetes mellitus in mice. *Nature* **351**, 542.

Wegmann, D. R., Norbury-Glaser, M., and Daniel, D. (1994). Insulin-specific T cells are a predominant component of islet infiltrates in pre-diabetic NOD mice. *Eur. J. Immunol.* **24**, 1853–1857.

Wicker, L. S., Clark, J., Fraser, H. I., Garner, V. E., Gonzalez-Munoz, A., Healy, B., Howlett, S., Hunter, K., Rainbow, D., Rosa, R. L., Smink, L. J., Todd, J. A., *et al.* (2005). Type 1 diabetes genes and pathways shared by humans and NOD mice. *J. Autoimmun.* **25**(Suppl.), 29.

Zhang, L., Nakayama, M., and Eisenbarth, G. S. (2008). *Curr. Opin. Immunol.* **20**, 111–118.

CHAPTER 2

CD3 Antibodies as Unique Tools to Restore Self-Tolerance in Established Autoimmunity: Their Mode of Action and Clinical Application in Type 1 Diabetes

Sylvaine You,[*,†] **Sophie Candon,**[*,†] **Chantal Kuhn,**[*,†] **Jean-François Bach,**[*,†] and **Lucienne Chatenoud**[*,†]

Contents		
	1. Introduction	14
	2. CD3 Antibodies as Promising Tolerance-Promoting Tools: The Proof of Concept in T1D	16
	3. The Clinical Results in T1D: More than 25 Years of Effort	18
	3.1. The initial data in clinical transplantation	18
	3.2. Back to the bench finding good reasons to proceed	19
	4. Why is The Effect of CD3-Antibody Based Therapy Antigen Specific and Long Lasting?	21
	5. Building on the Present Clinical Experience to Improve CD3 Antibody Therapy in T1D	26
	6. Conclusions	30
	References	30

[*] Université Paris Descartes, 75015 Paris, France
[†] INSERM, Unité 580, 75015 Paris, France

Advances in Immunology, Volume 100
ISSN 0065-2776, DOI: 10.1016/S0065-2776(08)00802-X

© 2008 Elsevier Inc.
All rights reserved.

1. INTRODUCTION

It was only during the 1980s that type 1 diabetes (T1D) has been clearly identified as a polygenic autoimmune disease due to the selective destruction of insulin secreting β cells within pancreatic islets of Langerhans by autoreactive T lymphocytes (Bach, 1994). The diabetological community rapidly reacted to this important discovery, concentrating efforts to approach first the major issue of the early diagnosis of the immunological disease and secondly, to devise immune-based therapeutic strategies to delay and/or prevent disease progression. As compared to other autoimmune diseases the situation in T1D was in fact quite unique due to the availability of spontaneous experimental models of the disease, the Bio Breeding (BB) rat and the nonobese diabetic (NOD) mouse (Bach, 1994; Crisa *et al.*, 1992; Makino *et al.*, 1980), which allowed dissecting the various stages of disease progression. It clearly appeared from the study of these models that the advent of an abnormal metabolic control, as assessed by a hyperglycemia and a glycosuria, considered as the hallmarks of T1D clinical diagnosis, was preceded by a long phase defined as *prediabetes* during which the β cell autoantigen-specific response silently, yet progressively, develops. In rodents prediabetes can be directly visualized by the histopathological analysis of the target organ. Thus, progressive infiltration of the islets of Langerhans by mononuclear cells is observed which, interestingly enough, is initially confined to the periphery of the islets (peri-insulitis) without any sign of active destruction of insulin secreting β cells (Bach, 1994; Bach and Chatenoud, 2001; Crisa *et al.*, 1992; Katz *et al.*, 1993; Makino *et al.*, 1980). As disease progresses the infiltrating cells invade the islets i.e., aggressive insulitis and β cell destruction is triggered. In the clinic where access to the target organ is exceptional, epidemiological studies conducted within families of T1D patients sequentially monitored for the advent of anti-islet cell autoantibodies, were instrumental to approach the natural history of prediabetes. It has been very well established that although anti-islet cell autoantibodies are nonpathogenic, they represent invaluable markers of β cell destruction. Thus, siblings of T1D patients stably presenting at least two among the major specificities of anti-islet cell autoantibodies (anti-insulin, anti-GAD, and anti-IA2) have a 60–80% risk of developing overt hyperglycemia within 5–7 years, respectively (Bingley and Gale, 2006).

Based on this knowledge and with the consciousness that conventional chronic insulin therapy is only a replacement strategy far from ideal due to major constraints (multiple daily parenteral administration, risk of hypoglycemia) and insufficient effectiveness in preventing severe degenerative complications, quite rapidly the first immunointervention approaches were attempted. Here again, the diabetological community considered as a high priority to conduct innovative randomized trials, very often placebo controlled, whose rationale was in the direct continuation of preclinical data recovered from animal studies. It is important to highlight the major breakthrough cyclosporin trials represented in this context (Feutren et al., 1986). They demonstrated for the first time that a T cell directed immunointervention could reverse established disease, totally disproving the prevailing dogma that almost all β cells have been destroyed at the time overt hyperglycemia. It is now well established from both experimental and clinical data that about 30% of potentially functional β cells are still present at the time of hyperglycemia (Sreenan et al., 1999). However, it has also been established that the immune-mediated inflammation severely impairs the insulin-secreting ability of remnant β cells, an effect which is fully reversible if rapid clearance of insulitis is achieved (Strandell et al., 1990). However, the activity of cyclosporin was targeted to intracellular processes essentially driving immunosuppression which is nonspecific, unrelated to the autoantigens involved, a major drawback which, in addition to its nephrotoxic potential, prevented cyclosporin from becoming an established therapy for T1D. The drug was relatively ineffective over the long term with the invariable recurrence of the pathogenic immune process once it was withdrawn, so necessitating indefinite administration which was an impracticable approach. In fact, T1D being a clinical condition mainly affecting children and young adults, the risk/benefit ratio of any therapeutics is thought-worthy.

Given these results, the second important step was to focus on more ambitious forms of immunotherapy not aiming at immunosuppression but at restoring self tolerance, a condition operationally defined as the inhibition of autoantigen-specific pathogenic immune responses in the absence of chronic immunosuppression (Bach and Chatenoud, 2001). Based on the experimental data showing the remarkable effect of autoantigen(s) administration in preventing disease progression, the Diabetes Prevention Trial DPT1 was launched. Results from this large controlled trial where insulin was delivered orally or subcutaneously to prediabetic children were mostly negative (Diabetes Prevention Trial-Type 1 Diabetes Study Group, 2002). Problems related to the dose used, patients' selection and, perhaps more importantly, the timing of administration were invoked to explain the results.

The key question is whether therapeutic self tolerance induction always needs the administration of exogenous autoantigen(s). Here is another dogma against which compelling evidence has been accumulated from both the autoimmunity and the transplantation field. Inasmuch as the autoantigen is present in the host, and we previously discussed that this is indeed the case in T1D even at the time of established hyperglycemia, drugs endowed with tolerogenic properties are available. In particular, it is well established that biological immunosuppressants notably anti-lymphocyte antibodies, in spite of their intrinsic lack of antigen specificity, are able to induce long-standing antigen-specific tolerance (Maki et al., 1981, 1992; Wood et al., 1971). This was particularly well demonstrated in experimental models for polyclonal anti-lymphocyte sera, monoclonal antibodies, and fusion proteins targeting key functionally relevant T cell receptors and/or signaling pathways (Pearson et al., 1992; Qin et al., 1993; Shizuru et al., 1987; Wekerle et al., 2002; Wofsy et al., 1985). Our aim here is to recount the long path which recently led us to surprising and highly encouraging results that stem from the transfer to the clinic of this type of immunointervention namely, a short treatment of newly diagnosed T1D patients with a CD3 monoclonal antibody, a strategy we established in the NOD mouse model more than 10 years back (Chatenoud, 2003; Chatenoud and Bluestone, 2007).

2. CD3 ANTIBODIES AS PROMISING TOLERANCE-PROMOTING TOOLS: THE PROOF OF CONCEPT IN T1D

Two clinical trials were initiated in early 2000 using two different antibodies to human CD3 (Bolt et al., 1993; Xu et al., 2000). A Phase I/II trial, conducted in the USA by the group of K. Herold used the humanized Fc-mutated antibody hOKT3γ1 (Ala-Ala) administered alone to treat children and young adult patients presenting T1D for 2 weeks at the time of disease onset (Herold et al., 2002, 2005). Evidence argued for a positive effect of the treatment in halting the progression of disease for several months, upto 36 months in some patients (Herold et al., 2005).

In Europe, we conducted a multinational, multicentre Phase II placebo-controlled clinical trial using the humanized Fc-mutated, aglycosylated ChAglyCD3 antibody. A total of 80 patients presenting new onset T1D receiving insulin treatment for not more than 4 weeks were randomized to receive a short 6-day treatment with 8 mg of ChAglyCD3 (40 patients) or placebo (40 patients). In this trial only adult patients were included. As already reported the antibody very efficiently preserved β-cell function, maintaining significantly higher levels of endogenous insulin secretion as compared to placebo-treated patients at 6, 12, and even 18 months after treatment. This was directly assessed by monitoring C-peptide levels after controlled intravenous glucose stimulation (Keymeulen et al., 2005). This effect translated into a very significant decrease in the patients' insulin needs during the same study period (Fig. 2.1) (Keymeulen et al., 2005). The multivariate analysis disclosed that one major predictive response factor was the functioning β cell mass present when starting the treatment. Patients were thus divided in two groups according to the median glucose-stimulated C-peptide levels at day zero and the most impressive effect in terms of β-cell mass preservation was observed in patients showing the higher stimulated C-peptide

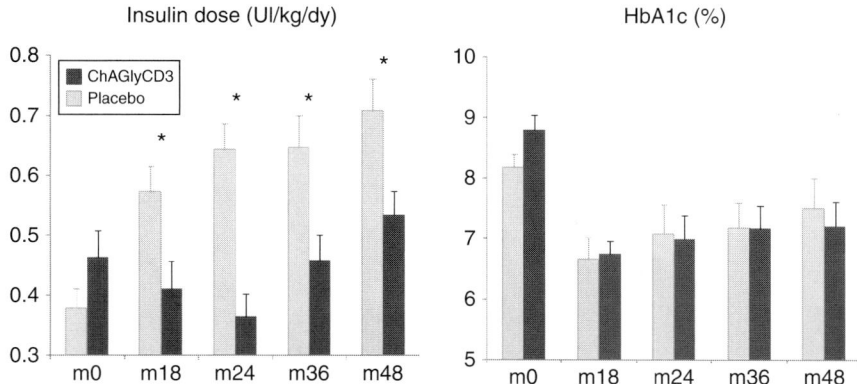

FIGURE 2.1 Long-term effect of CD3 antibody therapy in diabetic patients. The histograms represent the mean ± SEM daily insulin doses (left panel) and HbA1c percentages (right panel) in CD3 antibody- (clear bars) and placebo-treated (dark bars) patients before CD3 therapy and at month 18, 24, 36, and 48 following treatment. Statistically significant difference in mean values between the two groups is indicated as an asterisk ($p < 0.05$).

levels at the time of inclusion. Perhaps more impressively, more than 75% of the subset of patients with the high β-cell mass showed very low insulin needs, ≤ 0.25 U/kg/day at 18 months, a dose requirement compatible with clinical insulin independency (Keymeulen et al., 2005).

The study has been extended and the data from the 4-year follow-up are now available from a total of 64 patients who accepted to participate. Results showed a remarkably sustained effect as the significant decrease in insulin needs was observed at 48 months of follow-up in the intent-to-treat population. It is important to emphasize that at any of the time points did HbA1c levels differ between antibody- and placebo-treated patients, well indicating that the difference in insulin needs was not the consequence of an inadequate adaptation of the exogenous insulin treatment in either group (Fig. 2.1).

Side effects were overall minor; they were observed only during the first days to weeks from treatment and did not impede patients' recruitment.

Minor acute side effects linked to a limited, cytokine release were observed after the first infusions that, however, could be managed when needed with adequate palliative treatments (Keymeulen et al., 2005).

Transient reactivation of Epstein Barr virus (EBV) was also observed as assessed by an increase in numbers of EBV copies measured in peripheral blood mononuclear cells using quantitative polymerase chain reaction (PCR) for 10–20 days after the first injection (Keymeulen et al., 2005). Within 1–3 weeks, in all patients, the number of EBV copies returned to normal baseline pre-treatment levels. Concomitantly, an efficient humoral and cellular immune response specific to EBV developed that was comparable in intensity and kinetics to that observed in normal subjects following infectious mononucleosis. This effective anti-EBV response is for obvious reasons very important in terms of safety. In addition, it suggests that the antibody treatment affects the autoimmune reaction but does not alter immune responses to unrelated exogenous antigens, an observation compatible with the induction by the CD3 antibody of a state of operational immune tolerance.

3. THE CLINICAL RESULTS IN T1D: MORE THAN 25 YEARS OF EFFORT

3.1. The initial data in clinical transplantation

The story of CD3 antibodies developed based on paradoxes and, especially in their beginning, a large dose of empiricism. They were the first therapeutic monoclonal antibodies that entered the clinical arena in 1981, when the relevance, not to say the molecular complexity, of the target molecule was totally unknown. OKT3 (Muromomab®), a mouse IgG2a (Kung *et al.*, 1979), was injected to the first patient to reverse an ongoing renal allograft rejection episode (Cosimi *et al.*, 1981a) on the sole basis that *in vitro* it behaved similarly to polyclonal anti-lymphocyte antibodies, which use was already at that time well established in clinical transplantation. OKT3 stained all mature T cells and was a potent inhibitor of both proliferative response and generation of alloreactive cytotoxic effectors in mixed lymphocyte cultures. Amazingly, only a few years later the fine structure of the CD3 complex was unravelled showing its tight association to the major T lymphocyte receptor, the one involved in antigen recognition (Clevers *et al.*, 1988; Davis and Chien, 1999). Anti-T cell monoclonal antibodies are all tightly species-specific and CD3 antibodies do not escape the rule. All available monoclonals to human CD3 do not even cross-react with lymphocytes from common nonhuman primates such as *Rhesus* or *Cynomolgus*; they do recognize chimpanzee cells. This explains that in those early days OKT3 was not subjected to conventional toxicological analysis, which was particularly fortunate. In fact, as all first generation CD3 antibodies, OKT3 was a potent T cell mitogen (Van Wauwe *et al.*, 1980). Following the first injection, it induced the release of various cytokines, mostly T cell derived including TNF, IL-6, IFNγ, and IL-10. Although self limited (lasting just a few hours, only after the first and not the subsequent injections), this cytokine release elicited a "flu-like" syndrome which was life-threatening if high doses of the antibody were used (Abramowicz *et al.*, 1989; Chatenoud *et al.*, 1991).

This explains that the risk would have been high for this drug to be discarded from further development following conventional *in vivo* toxicological studies.

During the 1980s a series of controlled studies clearly demonstrated that OKT3 was a potent immunosuppressant very efficient at both reversing and preventing acute organ allograft rejection episodes (Cosimi *et al.*, 1981b; Ortho Multicenter Transplant Study Group, 1985), an indication for which this antibody was rapidly licensed worldwide.

Over the last 10 years, as other better tolerated biological immunosuppressants developed, the use of OKT3 was almost completely abandoned in transplantation, essentially because of its cytokine releasing potential (Abramowicz *et al.*, 1989; Chatenoud, 2003; Chatenoud *et al.*, 1986, 1989, 1990; Cosimi, 1987; Eason and Cosimi, 1999).

3.2. Back to the bench finding good reasons to proceed

The experimental work conducted in different rat and mouse models suggested that more than simply depressing all immune responses, CD3 antibodies were tolerogenic; they induced immune tolerance to both tissue alloantigens and autoantigens (Hayward and Shreiber, 1989; Nicolls *et al.*, 1993; Plain *et al.*, 1999) and, perhaps more impressively, could restore self-tolerance in established autoimmunity (Belghith *et al.*, 2003; Chatenoud, 2003; Chatenoud *et al.*, 1994, 1997).

The results of Nicolls *et al.* in a rat transplant model showed that a short CD3 antibody treatment induced permanent engraftment of fully mismatched vascularized heart allografts. Recipients were indeed tolerant since they accepted secondary donor-matched skin grafts while third party skin allografts were normally rejected (Nicolls *et al.*, 1993; Plain *et al.*, 1999). These results were interesting but in essence did not significantly differ from those reported with other tolerogenic anti-T cell monoclonal antibodies, such as anti-CD4 administered alone or in combination with anti-CD8, anti-CD40 Ligand, or some fusion proteins such as CTLA4-Ig (Pearson *et al.*, 1992; Qin *et al.*, 1993; Shizuru *et al.*, 1987; Wekerle *et al.*, 2002; Wofsy *et al.*, 1985). At variance, in autoimmunity CD3 antibodies expressed the unique ability to restore self-tolerance if administered when disease was established. In other words, they could "reprogram" the immune system in a primed host, an effect that has not been observed with any of the biological agents previously mentioned. This was first reported by our group in T1D using the NOD mouse model. A very short treatment with CD3 antibodies, only five consecutive days at low doses, when applied at the time of established disease i.e., early onset of hyperglycemia, could induce long-lasting remission of T1D (Chatenoud, 2003; Chatenoud *et al.*, 1994, 1997). The formal proof that this effect was due to restoration of self-tolerance to β-cell autoantigens came from *in vivo* experiments. Thus, syngeneic islet grafts (that, when implanted into untreated diabetic NOD recipients, are rapidly destroyed due to relapse of the autoimmune attack) survived indefinitely in CD3 antibody-treated mice as compared to fully mismatched skin allografts which were normally rejected (Chatenoud et al., 1994). In marked contrast with its remarkable therapeutic activity in established disease, the same treatment was totally ineffective to prevent T1D, when applied to young prediabetic NOD mice (4–8 weeks of age) which exhibit, as we described above, nondestructive benign insulitis. This observation was also surprising as it completely distinguished CD3 antibodies from the vast majority of other biological agents used in NOD mice that could prevent disease but were totally ineffective to reversing established T1D. The same type of results were reported using CD3 antibodies in another autoimmune disease namely, experimental allergic encephalomyelitis (EAE). Kohm et al. reported that CD3 antibodies were highly effective at treating ongoing proteolipid protein (PLP)-induced EAE in SJL mice, as assessed by acceleration of disease remission and prevention of further relapses (Kohm et al., 2005). More recently, comparable results were reported by Perruche et al. in both myelin oligodendrocyte glycoprotein (MOG)- and PLP-induced EAE (Perruche et al., 2008). Interesting data were also reported in 2,4,6-trinitrophenol-conjugated keyhole limpet hemocyanin (TNP-KLH)-induced colitis as a single CD3 antibody injection fully inhibited disease development (Ludviksson et al., 1997). In all these models CD3 antibodies were delivered through parenteral route (i.v.). To be complete, although the mechanisms underlying the therapeutic effect may differ, it is interesting to quote the puzzling data reported by the group of H. Weiner showing that oral delivery of CD3 antibody could both prevent and treat PLP-induced EAE in SJL mice and MOG-induced EAE in NOD mice (Ochi et al., 2006).

It is based on these data that CD3 antibodies were back in the clinical arena as engineered much better tolerated humanized Fc-mutated monoclonal antibodies (Alegre et al., 1994; Bolt et al., 1993) both in transplantation (Hering et al., 2004) and autoimmunity (Herold et al., 2002, 2005; Keymeulen et al., 2005; Utset et al., 2002). In fact, the mitogenic activity of CD3 antibodies is monocyte/macrophage-dependent and relies on the capacity of their Fc portion to bind Fc receptors (FcR). Thus, the mitogenic response significantly varies with the antibody isotype (i.e., murine IgA CD3 antibodies are far less mitogenic in humans (Parlevliet et al., 1994)) and F(ab')2 fragments, lacking the Fc portion, are nonmitogenic (Chatenoud et al., 1994, 1997; Hirsch et al., 1990; Van Lier et al., 1987). In all the experimental studies described above both mitogenic and nonmitogenic F(ab')2 fragments exhibited the same tolerogenic ability. This was the rationale for designing engineered FcR nonbinding humanized CD3-specific antibodies (Alegre et al., 1994; Bolt et al., 1993; Chatenoud et al., 1997;

Herold *et al.*, 1992; Hughes *et al.*, 1994; Johnson *et al.*, 1995). It is important to emphasize that FcR nonbinding CD3 antibodies still trigger partial signaling both *in vitro* and *in vivo*, an effect that is, as we shall discuss below, essential for their therapeutic activity (Chatenoud *et al.*, 1997; Smith *et al.*, 1997, 1998).

4. WHY IS THE EFFECT OF CD3-ANTIBODY BASED THERAPY ANTIGEN SPECIFIC AND LONG LASTING?

Relevant to this discussion are two main characteristics of the tolerogenic capacity of CD3 antibodies. First, the therapy is effective only in the context of a primed and ongoing immune response namely, when the autoimmune disease is overt. Thus, as we mentioned above, to promote long-term disease remission both in T1D in the NOD mouse and in the EAE model, the CD3 antibody therapy must be started once the effector stage of the autoimmune response is established. The therapy is still effective once the target tissue has been damaged to a significant degree. Second, CD3 antibody-mediated tolerance is completely blocked if a calcineurin inhibitor such as cyclosporin is administered concomitantly. As a whole, these two observations suggest first, that locally activated autoreactive T cell effectors are privileged targets of this therapy and secondly, that effective transduction of TCR/CD3 derived signals (i.e., signal 1) is an essential prerequisite for efficacy. One may thus hypothesize that although CD3 is expressed on all T lymphocytes, CD3 antibodies will trigger differential signaling depending on the functional capacity of the T cell target, i.e., a naïve T cell, an activated or a memory T cell, a regulatory T cell.

Experimental and clinical data have established that CD3 antibodies, even under the form of FcR nonbinding antibodies, promote some depletion which is however quantitatively limited (about 25% of all TCR/CD3 expressing T cells). The fact that F(ab')2 fragments mediate this effect points to the induction of apoptosis secondary to a direct antibody signaling effect or through redirected T cell lysis (Wong and Colvin, 1991). Data also support the conclusion that this depletion essentially involves activated effector T cells rather than regulatory T cells (Chatenoud *et al.*, 1994; Hirsch *et al.*, 1988). Thus, *in vitro* human activated T cells were shown to be more susceptible than resting ones to FcR nonbinding CD3 antibody-mediated apoptosis (Carpenter *et al.*, 2000; Wesselborg *et al.*, 1993). In fact, effector T cell death is most dramatic in the tissular target of the autoimmune process where the density of activated T cells is the highest. In contrast, CD4+CD25+FoxP3+(forkhead box P3) regulatory T cells (Treg) appear more resistant to CD3 antibody induced apoptosis compared with recently activated effector and naïve T cells (Tang and Bluestone, unpublished observations).

The other direct effect of CD3 antibody on their targets during treatment is antigenic modulation of the TCR/CD3 receptor. Antigenic modulation is the internalization and/or the shedding of the CD3 antibody/receptor complex from the cell surface, which is reversible within a few hours (8–12) once the antibody is cleared from the environment (Chatenoud *et al.*, 1982). TCR/CD3⁻ T cells are transiently blind to antigen and unresponsive to stimulation. In addition, the loss of TCR/CD3 expression on T cells infiltrating the target organ will, for obvious reasons, induce/facilitate their egress and redistribution in various other lymphocyte compartments. Altogether the privileged apoptosis of activated T cells and their migration out from the target organ are well in keeping with the rapid and complete clearance of insulitis observed in diabetic NOD mice by the end of the antibody treatment (Chatenoud *et al.*, 1994, 1997). The re-establishment of a noninflamed environment allows the remaining viable β-cells to recover their insulin-producing capacity thus explaining the rapid remission of T1D. Interestingly, *in vitro* data suggested that the rate of CD3 antibody-induced apoptosis was inversely correlated with that of antigenic modulation suggesting, as intuitively one would expect, that persistence of some level of CD3/TCR complex on the cell surface is required for sustained signaling and subsequent apoptosis (Carpenter *et al.*, 2000). This may be relevant in terms of antibody dosing, low doses being more prone to drive apoptosis, a hypothesis that still needs experimental confirmation. In the T1D model, at least in our hands, clearance of insulitis that drives the rapid initial reversal of established disease through physical eradication and functional inactivation of pathogenic T cells, requires repeated injections of CD3 antibodies (total of five doses using 25 μg/injection i.v. or total of three using 50 μg/injection).

However, at this point it is fair to admit that, although biologically relevant, none of these two phenomena, T cell depletion and antigenic modulation, which are transient, only operate during the short course treatment and are both rapidly reversible after the end of treatment, do explain the antigen-specific unresponsiveness which is sustained and prevents further relapses of the autoimmune process.

Peripheral tolerance mechanisms are operational during this phase, which, at least *in vivo*, appears to rely on the presence of a T cell-mediated active or dominant tolerance mainly involving, as we shall discuss in a moment, TGF-β-dependent adaptive Tregs. Data from *in vitro* experiments have suggested that unresponsiveness could also be induced through anergy. Thus, signaling triggered by FcR-nonbinding CD3 antibodies on Th1 T cell clones leads to a lack of phosphorylation of recruited ZAP70 correlating with aborted T-cell activation as assessed by deficient calcium influx, blockade of interleukin-2 (IL-2) production

(Smith *et al.*, 1997). In contrast, Th2 clones or polyclonal IL-4-secreting T cells exposed to the same stimulation exhibited identical signaling defects but remained responsive in terms of proliferation capacity (Smith *et al.*, 1998). These data point again at the nature of the T cell (activated vs. resting T cell, effector vs. regulatory T cell) as a determining factor for the "quality" of the CD3 antibody-mediated signal and the resulting outcome (Chatenoud and Bluestone, 2007). Our data *in vivo* do not point to anergy as a mechanism explaining the long-term therapeutic effect of the antibody. The problem may be however that in the "polyclonal" situation of wild type NOD mice anergy, which by definition implicates autoantigen-specific T cells, is not easy, not to say impossible, to assess. We are presently using more contrived, yet more amenable to antigen-specific cell tracking, transgenic models to directly address this issue.

The central question is which cellular and molecular mechanism(s) drive the long standing T cell-mediated immunoregulation and what is the causal link between the specific CD3 antibody impact on T cells during treatment (i.e., apoptosis and antigenic modulation), which we could define as the tolerance "induction" phase, and these immune mechanisms operating over the long term in absence of the antibody, which we could define as the tolerance "maintenance" phase. Based on our data, we would like to propose TGF-β production as the major causal link. Indeed, after CD3 antibody treatment, important amounts of soluble and membrane-bound TGF-β are produced by various cell types, including both CD4+CD25+ and CD4+CD25- T lymphocytes as well as non-T cells, starting as early as 1 week following therapy and lasting for several weeks (Belghith *et al.*, 2003). Furthermore, TGF-β blockade, through administration of specific neutralizing antibodies, fully abrogates the CD3 antibody tolerogenic effect (Belghith *et al.*, 2003) (Fig. 2.2). Other immunoregulatory cytokines do not exhibit the same pattern. For instance, a short burst of IL-4 production, was also observed after the end of treatment, suggesting transient TH2-cell polarization. However, this did not appear to impact the tolerogenic effect of the antibody since CD3 antibody-induced disease remission was observed in NOD/*Il4*$^{-/-}$ mice as readily as in wild-type NOD mice (You *et al.*, 2006).

It is tempting to link these observations to the recent data reported by the group of W.J. Chen (Chen *et al.*, 2001; Perruche *et al.*, 2008), demonstrating first, that local phagocytes (macrophages or immature dendritic cells) engulfing apoptotic cells are major sources of TGF-β and secondly, that in the EAE model the tolerogenic effect of CD3 antibody is tightly dependent on the presence of phagocytes. If phagocytes are depleted using clodronate-loaded liposomes, the therapeutic effect is lost.

In our working model (Fig. 2.3), we would like to propose that the CD3 antibody-mediated maintenance tolerance phase is triggered during the few days of antibody treatment by a massive local production of TGF-β.

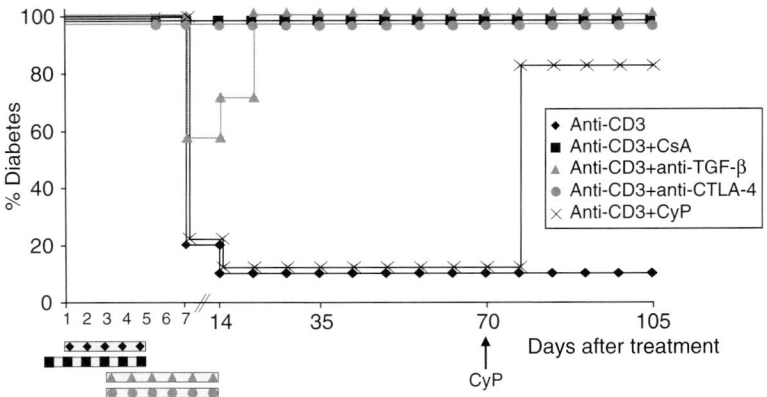

FIGURE 2.2 Reversal of CD3-specific antibody tolerogenic effect in the NOD mouse model. Overtly diabetic NOD mice were treated with CD3-specific antibodies for five consecutive days (D1–D5) alone or in combination with cyclosporin (20 mg/kg on D1–D5, red square), anti-TGF-β antibodies (1 mg on D3, D5, D8, D10, D12, D14, yellow triangle), anti-CTLA-4 antibodies (1 mg on D3, D5, D8, D10, D12, D14, green circle), or cyclophosphamide (200 mg/kg on D70). (See Plate 1 in Color Plate Section.)

This TGF-β would essentially be released by phagocytes engulfing activated/pathogenic effector T cells which undergo CD3 antibody-mediated apoptosis (Fig. 2.3). It will be relevant to explore further whether local Tregs, mostly of the adaptive type, which are more resistant to CD3 antibody-mediated apoptosis and by definition should undergo antigenic modulation as all other T cells, also participate to the TGF-β production in this early phase.

In support of a central action for local TGF-β production is the compelling evidence in the literature pointing to the privileged role of this cytokine to convert a pro-inflammatory environment into a noninflammatory and tolerogenic one (Li et al., 2006). High concentrations of TGF-β favor a decrease in the expression of MHC and costimulatory molecules such as CD80 and CD86 on antigen presenting cells (Li et al., 2006) while they promote the upregulation of inhibitory receptor ligands such as PDL-1 (programmed cell death ligand 1) or ICOS ligand (Rutella et al., 2006; Steinman et al., 2003). In turn, these antigen presenting cells are known to favor the induction and/or expansion of Tregs (Rutella et al., 2006; Yamazaki et al., 2007).

Although they need confirmation, we collected preliminary data showing increased expression of PDL-1 and ICOS ligand on dendritic cells recovered from pancreatic lymph nodes of CD3 antibody-treated NOD mice (unpublished data). The importance of this overexpression of PDL-1 is well in keeping with the data from the group of J.A. Bluestone showing that blockade of this pathway breaks CD3 antibody-induced tolerance (Fife *et al.*, 2006). In addition, we have well established that active and transferable tolerance mediated by a specialized subset of adaptive Tregs effectively controlling diabetogenic T cell effectors is operational in NOD mice showing restoration of self-tolerance following CD3 antibody therapy. In fact, within a few days after the end of treatment, following clearance of the antibody, insulitis recurs that is, as in prediabetic animals, noninvasive and nonactively destructing, yet indicating that the tolerant hosts harbor autoreactive T cells and that β-cell autoantigens are effectively presented (Chatenoud *et al.*, 1994, 1997). In parallel, CD4+CD25low Tregs expand in CD3 antibody-treated tolerant mice (Belghith *et al.*, 2003; You *et al.*, 2006, 2007). Proportions of these Tregs increase as early as 1 week following treatment, they express FoxP3 and display suppressive properties that are TGF-β-dependent both *in vitro* (in conventional coculture assays) and *in vivo* (if cotransferred with diabetogenic T cells into immunoincompetent syngeneic recipients). Further pointing to the adaptive nature of these Tregs are the data from CD3 antibody-treated diabetic NOD/cd28$^{-/-}$ (which are devoid of thymic-derived natural Tregs (Salomon *et al.*, 2000)) showing long-term disease remission just as wild-type NOD that correlates with the induction in the periphery of CD4+CD25^{low}FoxP3$^+$ Tregs that here again are TGF-β-dependent (You *et al.*, 2007).

The last but not the least question to address is whether, and if yes how, all this is sufficient to explain a long-term effect. Data from the fields of transplantation, tumor immunology and mucosal immunity may be helpful in bringing the puzzle together. In other words, may pancreatic islets and draining lymph nodes from NOD mice that are tolerant following CD3 antibody therapy acquire the immune status of a "privileged" tissue?

Immune privilege was originally defined as particular sites where allogeneic tissues such as the anterior chamber of the eye, the placenta, the testis, or the brain could be successfully transplanted without eliciting rejection. Immune privilege was initially considered to rely on specific anatomical features of these tissues affording them the capacity to "escape" inflammation. Over last years, it became clearer that immune privilege can be acquired locally in almost any tissue in response to an inflammatory insult and also following therapeutic manipulations aiming at tolerance induction (Mellor and Munn, 2008; Waldmann *et al.*, 2006). A prototypic example, having direct implications for our model, is that of organ allografts that are indefinitely accepted following a short treatment of the host with anti-T cell monoclonal antibodies to CD4 and CD8 that induce specific tolerance to the alloantigens (Cobbold *et al.*, 2006). In this model too TGF-β plays a major role. It orchestrates the interplay between tissue alloantigens which uptake by infiltrating dendritic cells in presence of Tregs drives the induction of anti-inflammatory cytokines and receptors mediating negative costimulatory signals that in turn modulate in a self-perpetuating manner both T cell and antigen presenting cell activity (Cobbold et al., 2006). In the model we propose (Fig. 2.3), the local noninflammatory milieu created during CD3 antibody therapy (the tolerance induction phase) drives high concentrations of TGF-β that "reprograms" the phenotypic and functional characteristics of local antigen presenting cells capturing β-cell autoantigens towards a tolerogenic/noninflammatory phenotype. These antigen presenting cells would be less prone to driving the proliferation/differentiation of autoantigen-specific pathogenic T while they would promote the expansion of pre-existing adaptive CD4+CD25low autoantigen-specific regulatory T cells (Matsumura et al., 2007; Rutella et al., 2006) as well as the conversion, in presence of TGF-β, of CD4+CD25- T cells into adaptive Tregs (Luo et al., 2007).

The nature of the various agents able to break the CD3 antibody-induced tolerance argues in favor of the various steps involved in the proposed model (Figs. 2.2 and 2.3). We already mentioned the case of neutralizing antibodies to TGF-β. Cyclosporin which is a potent blocker of the CD3 antibody tolerogenic effect (Chatenoud et al., 1997) inhibits T cell apoptosis (Lenardo, 1991; Takahashi et al., 2004; Wells et al., 1999; Woodside et al., 2006; Yazdanbakhsh et al., 1995) as well as the generation and expansion of Tregs (Coenen et al., 2007; Zeiser et al., 2006). Cyclophosphamide, that also promotes rapid reversal of CD3 antibody-induced protection (Chatenoud et al., 1997), is an alkylating agent well known for its selective activity on Tregs (Askenase et al., 1975; Yasunami and Bach, 1988). The blocking effect of a neutralizing antibody to CTLA-4 may also, at least in part, be ascribed to an effect on Tregs which express this receptor (Belghith et al., 2003).

5. BUILDING ON THE PRESENT CLINICAL EXPERIENCE TO IMPROVE CD3 ANTIBODY THERAPY IN T1D

Although CD3 monoclonal antibodies presently used in the clinic exhibit dramatically reduced T-cell activation-induced side effects after the first dose as compared to FcR-binding CD3 antibodies such as OKT3, some degree of T-cell activation is still observed (Friend et al., 1999; Herold et al., 2002; Keymeulen et al., 2005; Woodle et al., 1999). This leads to minor symptoms that mainly include moderate fever, headaches, and self-limiting gastrointestinal manifestations, all of which are amenable to palliative treatment.

However, for obvious reasons, if this is intended to become an established therapy for T1D, and especially for pediatric patients, it will be important to further improve safety by fully preventing these symptoms. Other variants of CD3 antibodies could be developed devoid of the Fc

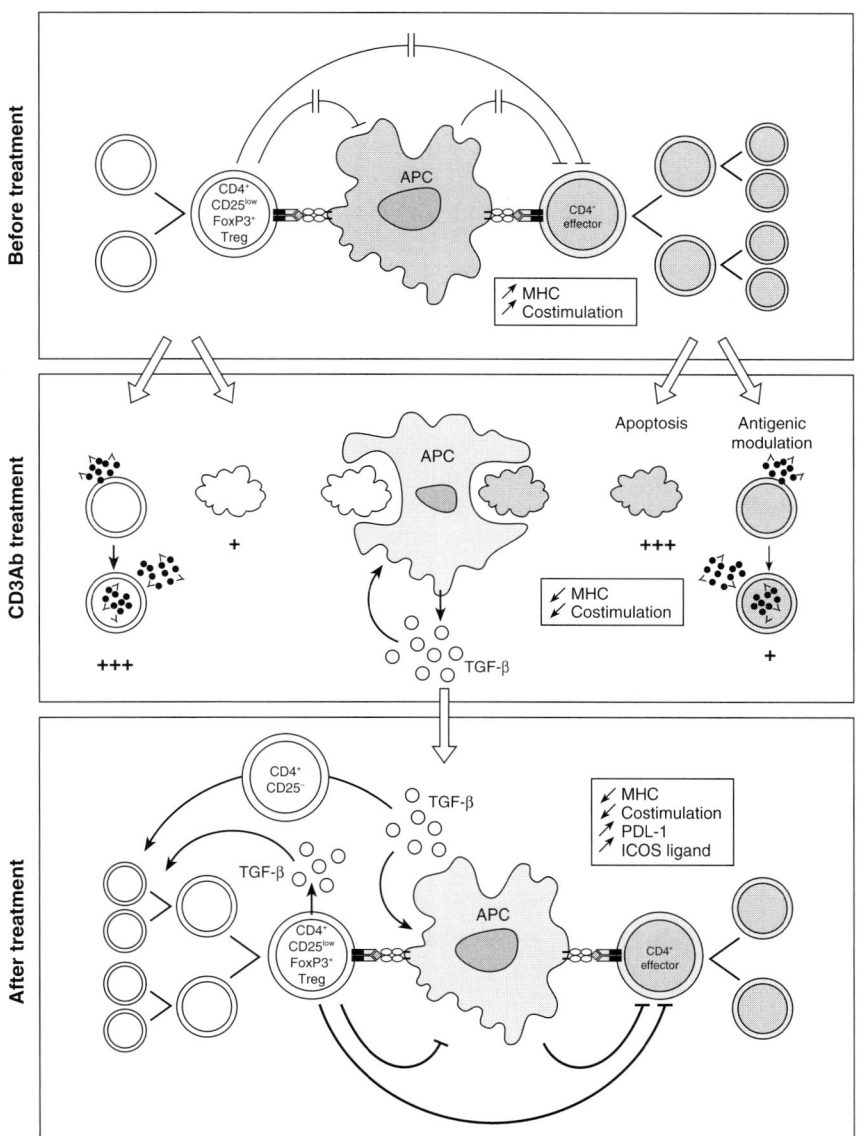

FIGURE 2.3 Mechanistic hypothesis to explain the tolerogenic effect of CD3 antibodies. (1) In draining lymph nodes, effector/pathogenic T cells are activated by their cognate autoantigens presented by professional antigen presenting cells (APCs) expressing the appropriate panel of costimulatory molecules and producing pro-inflammatory cytokines. These pathogenic T cells are resistant to the direct or indirect control exerted by CD4+ Foxp3+ regulatory T cells and actively proliferate thus contributing to the destructive autoimmune response. (2) Administration of CD3 antibodies induces a significant depletion of T cells and/or TCR/CD3 complex antigenic modulation in

portion (i.e., pegylated F(ab')2 type constructs) that could, as observed in the mouse models, completely avoid the cytokine release, yet keeping intact the signaling capacity. While we wait for the advent of these products, another possibility is to use pre-medications that proved successful at preventing the CD3 antibody-induced cytokine release. This was the case for instance of a single administration of TNF specific antibody administered one hour prior to OKT3 administration (Charpentier *et al.*, 1992; Ferran *et al.*, 1990, 1991, 1994).

In patients who had already experienced a primary infection with EBV viral re-activation was observed. In terms of safety, this viral reactivation was transient, did not involve viruses other than EBV, and never recurred in any of the patients who have had long-term follow-up (4–7 years) (our unpublished data). As previously discussed, in all patients EBV copies returned to normal baseline pre-treatment values due to the rapid development of an efficient EBV-specific cellular immune response including in particular, significant numbers of $CD8^+$ T cells directed to lytic cycle EBV peptides. Although the mechanisms responsible for this EBV reactivation are still ill defined, we collected some evidence to suggest that it may be dose related and dependent on the T cell activation exhibited by Fc-mutated antibodies. Thus, at least in theory effectively preventing the T cell activation would also positively impact on the EBV reactivation.

Immunization to the CD3 antibody is also a concern. Independently from the humanized antibody used, about 40% of patients develop antiidiotypic antibodies (Herold *et al.*, 2002, 2005; Keymeulen *et al.*, 2005). Despite their potential neutralizing capacity, anti-idiotypic antibodies did

---

lymphoid organs and in the infiltrated tissues. In the draining lymph nodes, depletion essentially involves activated effector T cells rather than regulatory T cells. During this phase, apoptotic T cells, and thus in particular autoantigen-specific pathogenic T cells, are engulfed and digested by phagocytes (macrophages and immature dendritic cells) that are present locally. These phagocytes produce large amount of TGF-β that creates a noninflammatory environment. (3) This local noninflammatory milieu influences the phenotypic and functional characteristics of APCs. Dendritic cells that present specific autoantigens express less co-stimulatory molecules but deliver more inhibitory signals involving molecules such as PDL-1 (programmed cell death ligand 1) or ICOS ligand. These dendritic cells can inhibit autoantigen-specific pathogenic T cells and promote, in the presence of TGF-β, the expansion of adaptive $CD4^+$ $Foxp3^+$ regulatory T cells. In addition, in the presence of such tolerogenic dendritic cells and TGF-β, naïve autoreactive $CD4^+$ $CD25^-$ T cells can be converted into adaptive Tregs. The CD3-induced Tregs cells themselves secrete TGF-β (as well as IL-10, our unpublished data) and can control directly or indirectly the diabetogenic effectors that also recover their sensitivity to regulation in this noninflammatory environment. Altogether, these
events contribute to the immunoregulatory process that perpetuates the long-term maintenance of self-tolerance.

not represent a problem when a single CD3 antibody course was applied, as they appeared by 2–3 weeks after the end of the treatment. However, they may represent a potential problem if repeated treatment is needed. Here again, data from murine models support the notion that decreasing the T cell activation capacity of the antibody will prevent immunization. Thus, in mice F(ab′)2 fragments of CD3 antibody were less immunogenic than the intact parental molecule (Hirsch *et al.*, 1990). Another possibility which proved effective with various monoclonal antibodies to block sensitization (i.e., humanized anti-TNF (Feldmann and Maini, 2001)) is to administer a short low dose course of conventional immunosuppressants (such as azathioprine, mycophenolate mofetil, methotrexate) in association to the first course of the CD3 antibody.

Coming now to the therapeutic activity on the disease, although CD3 antibody therapy was remarkable in reversing established diabetes progression, some variability was noted among individuals. As we discussed, the best response was observed in patients presenting a high functioning β-cell mass at the beginning of treatment (Keymeulen *et al.*, 2005). One possibility to improve the outcome in overt T1D is to associate to CD3 antibody treatment drugs that promote β-cell survival and growth, such as incretin hormones—glucagon-like peptide-1 (GLP1) or its related mimic, exendin-4 (Wajchenberg, 2007). These agents augment glucose-stimulated insulin release and may also increase β-cell replication, and stimulate β-cell neogenesis. Indeed, the combination of exendin-4 and CD3 monoclonal antibody therapy led to a more effective reversal of spontaneous diabetes in NOD mice by enhancing the insulin content of pancreatic β-cells as compared with treatment with either agent alone (Sherry *et al.*, 2007).

The other, nonmutually exclusive, yet more tempting, avenue is proposing to start the treatment much earlier. We discussed in fact that the autoimmune aggression begins months to years before the appearance of overt diabetes. Treating patients earlier in the course of the disease would certainly have a better chance to improve results. In other words, moving from therapy of T1D at the time of established hyperglycemia to therapy of prediabetes is certainly the next sensible step. However, in order to make this feasible improving the safety of the product is certainly an essential prerequisite as the patients' population will in this case mainly include young children.

Recurrence of autoimmunity is another factor which may shorten the effect of CD3 antibody therapy. In this case, repeated treatment may be the solution provided that caution is taken to prevent, as discussed above, the advent of anti-idiotypic antibodies after the first antibody course.

A more sophisticated and probably more effective approach for the future will probably be the combination of immunotherapies taking advantage of synergistic effects, which may allow decreasing the dosage of each individual drug used. Among the various options, one must certainly quote the administration of autoantigens. In this respect the recent report by Bresson *et al.* examining treatment with a CD3 antibody combined with intranasal proinsulin peptide are very encouraging (Bresson *et al.*, 2006). Combined targeting of B cells using the CD20 monoclonal antibody rituximab is also an appealing approach, the rationale being that in autoimmune diabetes, B cells are known to be major autoantigen presenting cells (Serreze and Silveira, 2003; Serreze *et al.*, 1996). The results of a randomized trial, presently conducted by TRIALNET in the USA using rituximab alone in new onset T1D patients will be available soon. Combination of CD3 antibody therapy with agents mediating costimulation blockade could also be considered, given the promising data reported, following administration of the CTLA-4 Ig fusion protein in psoriasis (Abrams *et al.*, 1999, 2000) and refractory rheumatoid arthritis (Genovese *et al.*, 2005; Kremer *et al.*, 2003, 2005; Mackie *et al.*, 2005).

6. CONCLUSIONS

As for a variety of autoimmune and allergic diseases the incidence of T1D has tremendously increased in industrialized countries over the last three decades (Bach, 2002). This is well-illustrated by the recent report in the finish pediatric population describing that the incidence of T1D is increasing even faster since 2000. The expectation is that the number of new cases diagnosed at or before 14 years of age will double in the next 15 years and the age of onset will be younger (0–4 years) (Harjutsalo *et al.*, 2008). There is an urgent need for establishing safe immunointervention procedures that may provide a real cure for the disease without necessitating prolonged exposure of patients to immunosuppressive drugs with all the attendant problems of recurrent infections and drug toxicity. At variance with conventional therapies, CD3 antibodies afford for the first time the possibility to induce long-term effects following a short administration. The obvious challenge in the next future will be to improve the conditions of their use by ameliorating the safety profile, better defining the subsets of patients who will benefit more from the treatment and envisaging which drug combinations may be the most profitable. All this will of course benefit from advances made in devising assays to detect directly pathogenic islet-specific T cells or to quantify β-cell mass using noninvasive methods that will be a major added value for the selection and the monitoring of patients.

REFERENCES

Abramowicz, D., Schandene, L., Goldman, M., Crusiaux, A., Vereerstraeten, P., De Pauw, L., Wybran, J., Kinnaert, P., Dupont, E., and Toussaint, C. (1989). Release of tumor necrosis factor, interleukin-2, and gamma-interferon in serum after injection of OKT3 monoclonal antibody in kidney transplant recipients. *Transplantation* **47**, 606.

Abrams, J. R., Lebwohl, M. G., Guzzo, C. A., Jegasothy, B. V., Goldfarb, M. T., Goffe, B. S., Menter, A., Lowe, N. J., Krueger, G., Brown, M. J., Weiner, R. S., Birkhofer, M. J., *et al.* (1999). CTLA4Ig-mediated blockade of T-cell costimulation in patients with psoriasis vulgaris. *J. Clin. Invest.* **103,** 1243.

Abrams, J. R., Kelley, S. L., Hayes, E., Kikuchi, T., Brown, M. J., Kang, S., Lebwohl, M. G., Guzzo, C. A., Jegasothy, B. V., Linsley, P. S., and Krueger, J. G. (2000). Blockade of T lymphocyte costimulation with cytotoxic T lymphocyte-associated antigen 4-immunoglobulin (CTLA4Ig) reverses the cellular pathology of psoriatic plaques, including the activation of keratinocytes, dendritic cells, and endothelial cells. *J. Exp. Med.* **192,** 681.

Alegre, M. L., Peterson, L. J., Xu, D., Sattar, H. A., Jeyarajah, D. R., Kowalkowski, K., Thistlethwaite, J. R., Zivin, R. A., Jolliffe, L., and Bluestone, J. A. (1994). A non-activating "humanized" anti-CD3 monoclonal antibody retains immunosuppressive properties *in vivo*. *Transplantation* **57,** 1537.

Askenase, P. W., Hayden, B. J., and Gershon, R. K. (1975). Augmentation of delayed-type hypersensitivity by doses of cyclophosphamide which do not affect antibody responses. *J. Exp. Med.* **141,** 697.

Bach, J. F. (1994). Insulin-dependent diabetes mellitus as an autoimmune disease. *Endocr. Rev.* **15,** 516.

Bach, J. F. (2002). The effect of infections on susceptibility to autoimmune and allergic diseases. *N. Engl. J. Med.* **347,** 911.

Bach, J. F., and Chatenoud, L. (2001). Tolerance to islet autoantigens in type 1 diabetes. *Annu. Rev. Immunol.* **19,** 131.

Belghith, M., Bluestone, J. A., Barriot, S., Megret, J., Bach, J. F., and Chatenoud, L. (2003). TGF-beta-dependent mechanisms mediate restoration of self-tolerance induced by antibodies to CD3 in overt autoimmune diabetes. *Nat. Med.* **9,** 1202.

Bingley, P. J., and Gale, E. A. (2006). Progression to type 1 diabetes in islet cell antibody-positive relatives in the European Nicotinamide Diabetes Intervention Trial: The role of additional immune, genetic and metabolic markers of risk. *Diabetologia* **49,** 881.

Bolt, S., Routledge, E., Lloyd, I., Chatenoud, L., Pope, H., Gorman, S. D., Clark, M., and Waldmann, H. (1993). The generation of a humanized, non-mitogenic CD3 monoclonal antibody which retains *in vitro* immunosuppressive properties. *Eur. J. Immunol.* **23,** 403.

Bresson, D., Togher, L., Rodrigo, E., Chen, Y., Bluestone, J. A., Herold, K. C., and von Herrath, M. (2006). Anti-CD3 and nasal proinsulin combination therapy enhances remission from recent-onset autoimmune diabetes by inducing Tregs. *J. Clin. Invest.* **116,** 1371.

Carpenter, P. A., Pavlovic, S., Tso, J. Y., Press, O. W., Gooley, T., Yu, X. Z., and Anasetti, C. (2000). Non-Fc receptor-binding humanized anti-CD3 antibodies induce apoptosis of activated human T cells. *J. Immunol.* **165,** 6205.

Charpentier, B., Hiesse, C., Lantz, O., Ferran, C., Stephens, S., O'shaugnessy, D., Bodmer, M., Benoit, G., Bach, J. F., and Chatenoud, L. (1992). Evidence that antihuman tumor necrosis factor monoclonal antibody prevents OKT3-induced acute syndrome. *Transplantation* **54,** 997.

Chatenoud, L. (2003). CD3-specific antibody-induced active tolerance: From bench to bedside. *Nat. Rev. Immunol.* **3,** 123.

Chatenoud, L., and Bluestone, J. A. (2007). CD3-specific antibodies : A portal to the treatment of autoimmunity. *Nat. Rev. Immunol.* **7,** 622.

Chatenoud, L., Baudrihaye, M. F., Kreis, H., Goldstein, G., Schindler, J., and Bach, J. F. (1982). Human *in vivo* antigenic modulation induced by the anti-T cell OKT3 monoclonal antibody. *Eur. J. Immunol.* **12,** 979.

Chatenoud, L., Baudrihaye, M. F., Chkoff, N., Kreis, H., Goldstein, G., and Bach, J. F. (1986). Restriction of the human *in vivo* immune response against the mouse monoclonal antibody OKT3. *J. Immunol.* **137,** 830.

Chatenoud, L., Ferran, C., Reuter, A., Legendre, C., Gevaert, Y., Kreis, H., Franchimont, P., and Bach, J. F. (1989). Systemic reaction to the anti-T-cell monoclonal antibody OKT3 in relation to serum levels of tumor necrosis factor and interferon-gamma. *N. Engl. J. Med.* **320,** 1420.

Chatenoud, L., Ferran, C., Legendre, C., Thouard, I., Merite, S., Reuter, A., Gevaert, Y., Kreis, H., Franchimont, P., and Bach, J. F. (1990). In vivo cell activation following OKT3 administration. Systemic cytokine release and modulation by corticosteroids. *Transplantation* **49,** 697.

Chatenoud, L., Legendre, C., Ferran, C., Bach, J. F., and Kreis, H. (1991). Corticosteroid inhibition of the OKT3-induced cytokine-related syndrome—dosage and kinetics prerequisites. *Transplantation* **51,** 334.

Chatenoud, L., Thervet, E., Primo, J., and Bach, J. F. (1994). Anti-CD3 antibody induces long-term remission of overt autoimmunity in nonobese diabetic mice. *Proc. Natl. Acad. Sci. USA* **91,** 123.

Chatenoud, L., Primo, J., and Bach, J. F. (1997). CD3 antibody-induced dominant self tolerance in overtly diabetic NOD mice. *J. Immunol.* **158,** 2947.

Chen, W., Frank, M. E., Jin, W., and Wahl, S. M. (2001). TGF-beta released by apoptotic T cells contributes to an immunosuppressive milieu. *Immunity* **14,** 715.

Clevers, H., Alarcon, B., Wileman, T., and Terhorst, C. (1988). The T cell receptor/CD3 complex: A dynamic protein ensemble. *Annu. Rev. Immunol.* **6,** 629.

Cobbold, S. P., Adams, E., Graca, L., Daley, S., Yates, S., Paterson, A., Robertson, N. J., Nolan, K. F., Fairchild, P. J., and Waldmann, H. (2006). Immune privilege induced by regulatory T cells in transplantation tolerance. *Immunol. Rev.* **213,** 239.

Coenen, J. J., Koenen, H. J., van Rijssen, E., Kasran, A., Boon, L., Hilbrands, L. B., and Joosten, I. (2007). Rapamycin, not cyclosporine, permits thymic generation and peripheral preservation of CD4+ CD25+ FoxP3+ T cells. *Bone Marrow Transplant.* **39,** 537.

Cosimi, A. B. (1987). Clinical development of orthoclone OKT3. *Transplant. Proc.* **19,** 7.

Cosimi, A. B., Burton, R. C., Colvin, R. B., Goldstein, G., Delmonico, F. L., Laquaglia, M. P., Tolkoff-rubin, N., Rubin, R. H., Herrin, J. T., and Russell, P. S. (1981a). Treatment of acute renal allograft rejection with OKT3 monoclonal antibody. *Transplantation* **32,** 535.

Cosimi, A. B., Colvin, R. B., Burton, R. C., Rubin, R. H., Goldstein, G., Kung, P. C., Hansen, W. P., Delmonico, F. L., and Russell, P. S. (1981b). Use of monoclonal antibodies to T-cell subsets for immunologic monitoring and treatment in recipients of renal allografts. *N. Engl. J. Med.* **305,** 308.

Crisa, L., Mordes, J. P., and Rossini, A. A. (1992). Autoimmune diabetes mellitus in the BB rat. *Diabetes Metab. Rev.* **8,** 4.

Davis, M. M., and Chien, Y. H. (1999). In "Fundamental Immunology" (W. Paul, Ed.), p. 341. Raven Press, New York.

Diabetes Prevention Trial-Type 1 Diabetes Study Group (2002). Effects of insulin in relatives of patients with type 1 diabetes mellitus. *N. Engl. J. Med.* **346,** 1685.

Eason, J. D., and Cosimi, A. B. (1999). In "Transplantation" (L. Ginns, , A. Cosimi, and , and P. Morris, Eds.), p. 196. Blackwell Science, Malden.

Feldmann, M., and Maini, T. (2001). Anti-TNF alpha therapy of rheumatoid arthritis: What have we learned? *Annu. Rev. Immunol.* **19,** 163.

Ferran, C., Dy, M., Merite, S., Sheehan, K., Schreiber, R., Leboulenger, F., Landais, P., Bluestone, J., Bach, J. F., and Chatenoud, L. (1990). Reduction of morbidity and cytokine release in anti-CD3 MoAb-treated mice by corticosteroids. *Transplantation* **50,** 642.

Ferran, C., Dy, M., Sheehan, K., Schreiber, R., Grau, G., Bluestone, J., Bach, J. F., and Chatenoud, L. (1991). Cascade modulation by anti-tumor necrosis factor monoclonal antibody of interferon-gamma, interleukin 3 and interleukin 6 release after triggering of the CD3/T cell receptor activation pathway. *Eur. J. Immunol.* **21,** 2349.

Ferran, C., Dautry, F., Merite, S., Sheehan, K., Schreiber, R., Grau, G., Bach, J. F., and Chatenoud, L. (1994). Anti-tumor necrosis factor modulates anti-CD3-triggered T cell cytokine gene expression *in vivo*. *J. Clin. Invest.* **93**, 2189.

Feutren, G., Papoz, L., Assan, R., Vialettes, B., Karsenty, G., Vexiau, P., Du Rostu, H., Rodier, M., Sirmai, J., Lallemand, A., and Bach, J. F. (1986). Cyclosporin increases the rate and length of remissions in insulin-dependent diabetes of recent onset. Results of a multicentre double-blind trial. *Lancet* **2**, 119.

Fife, B. T., Guleria, I., Gubbels Bupp, M., Eagar, T. N., Tang, Q., Bour-Jordan, H., Yagita, H., Azuma, M., Sayegh, M. H., and Bluestone, J. A. (2006). Insulin-induced remission in new-onset NOD mice is maintained by the PD-1-PD-L1 pathway. *J. Exp. Med.* **203**, 2737.

Friend, P. J., Hale, G., Chatenoud, L., Rebello, P., Bradley, J., Thiru, S., Phillips, J. M., and Waldmann, H. (1999). Phase I study of an engineered aglycosylated humanized CD3 antibody in renal transplant rejection. *Transplantation* **68**, 1632.

Genovese, M. C., Becker, J. C., Schiff, M., Luggen, M., Sherrer, Y., Kremer, J., Birbara, C., Box, J., Natarajan, K., Nuamah, I., Li, T., Aranda, R., *et al.* (2005). Abatacept for rheumatoid arthritis refractory to tumor necrosis factor alpha inhibition. *N. Engl. J. Med.* **353**, 1114.

Harjutsalo, V., Sjoberg, L., and Tuomilehto, J. (2008). Time trends in the incidence of type 1 diabetes in Finnish children: A cohort study. *Lancet* **371**, 1777.

Hayward, A. R., and Shreiber, M. (1989). Neonatal injection of CD3 antibody into nonobese diabetic mice reduces the incidence of insulitis and diabetes. *J. Immunol.* **143**, 1555.

Hering, B. J., Kandaswamy, R., Harmon, J. V., Ansite, J. D., Clemmings, S. M., Sakai, T., Paraskevas, S., Eckman, P. M., Sageshima, J., Nakano, M., Sawada, T., Matsumoto, I., *et al.* (2004). Transplantation of cultured islets from two-layer preserved pancreases in type 1 diabetes with anti-CD3 antibody. *Am. J. Transplant.* **4**, 390.

Herold, K. C., Bluestone, J. A., Montag, A. G., Parihar, A., Wiegner, A., Gress, R. E., and Hirsch, R. (1992). Prevention of autoimmune diabetes with nonactivating anti-CD3 monoclonal antibody. *Diabetes* **41**, 385.

Herold, K. C., Hagopian, W., Auger, J. A., Poumian Ruiz, E., Taylor, L., Donaldson, D., Gitelman, S. E., Harlan, D. M., Xu, D., Zivin, R. A., and Bluestone, J. A. (2002). Anti-CD3 monoclonal antibody in new-onset type 1 diabetes mellitus. *N. Engl. J. Med.* **346**, 1692.

Herold, K. C., Gitelman, S. E., Masharani, U., Hagopian, W., Bisikirska, B., Donaldson, D., Rother, K., Diamond, B., Harlan, D. M., and Bluestone, J. A. (2005). A single course of anti-CD3 monoclonal antibody hOKT3gamma1(Ala-Ala) results in improvement in C-peptide responses and clinical parameters for at least 2 years after onset of type 1 diabetes. *Diabetes* **54**, 1763.

Hirsch, R., Eckhaus, M., Auchincloss, H. J. R., Sachs, D. H., and Bluestone, J. A. (1988). Effects of *in vivo* administration of anti-T3 monoclonal antibody on T cell function in mice. I. Immunosuppression of transplantation responses. *J. Immunol.* **140**, 3766.

Hirsch, R., Bluestone, J. A., De Nenno, L., and Gress, R. E. (1990). Anti-CD3 F(ab')2 fragments are immunosuppressive *in vivo* without evoking either the strong humoral response or morbidity associated with whole mAb. *Transplantation* **49**, 1117.

Hughes, C., Wolos, J. A., Giannini, E. H., and Hirsch, R. (1994). Induction of T helper cell hyporesponsiveness in an experimental model of autoimmunity by using nonmitogenic anti-CD3 monoclonal antibody. *J. Immunol.* **153**, 3319.

Johnson, B. D., Mccabe, C., Hanke, C. A., and Truitt, R. L. (1995). Use of anti-CD3 epsilon F (ab')2 fragments *in vivo* to modulate graft-versus-host disease without loss of graft-versus-leukemia reactivity after MHC-matched bone marrow transplantation. *J. Immunol.* **154**, 5542.

Katz, J. D., Wang, B., Haskins, K., Benoist, C., and Mathis, D. (1993). Following a diabetogenic T cell from genesis through pathogenesis. *Cell* **74**, 1089.

Keymeulen, B., Vandemeulebroucke, E., Ziegler, A. G., Mathieu, C., Kaufman, L., Hale, G., Gorus, F., Goldman, M., Walter, M., Candon, S., Schandene, L., Crenier, L., *et al*. (2005). Insulin needs after CD3-antibody therapy in new-onset type 1 diabetes. *N. Engl. J. Med.* **352,** 2598.

Kohm, A. P., Williams, J. S., Bickford, A. L., McMahon, J. S., Chatenoud, L., Bach, J. F., Bluestone, J. A., and Miller, S. D. (2005). Treatment with nonmitogenic anti-CD3 monoclonal antibody induces CD4+ T cell unresponsiveness and functional reversal of established experimental autoimmune encephalomyelitis. *J. Immunol.* **174,** 4525.

Kremer, J. M., Westhovens, R., Leon, M., Di Giorgio, E., Alten, R., Steinfeld, S., Russell, A., Dougados, M., Emery, P., Nuamah, I. F., Williams, G. R., Becker, J. C., *et al*. (2003). Treatment of rheumatoid arthritis by selective inhibition of T-cell activation with fusion protein CTLA4Ig. *N. Engl. J. Med.* **349,** 1907.

Kremer, J. M., Dougados, M., Emery, P., Durez, P., Sibilia, J., Shergy, W., Steinfeld, S., Tindall, E., Becker, J. C., Li, T., Nuamah, I. F., Aranda, R., *et al*. (2005). Treatment of rheumatoid arthritis with the selective costimulation modulator abatacept: Twelve-month results of a phase iib, double-blind, randomized, placebo-controlled trial. *Arthritis Rheum.* **52,** 2263.

Kung, P., Goldstein, G., Reinherz, E. L., and Schlossman, S. F. (1979). Monoclonal antibodies defining distinctive human T cell surface antigens. *Science* **206,** 347.

Lenardo, M. J. (1991). Interleukin-2 programs mouse alpha beta T lymphocytes for apoptosis. *Nature* **353,** 858.

Li, M. O., Wan, Y. Y., Sanjabi, S., Robertson, A. K., and Flavell, R. A. (2006). Transforming growth factor-beta regulation of immune responses. *Annu. Rev. Immunol.* **24,** 99.

Ludviksson, B. R., Ehrhardt, R. O., and Strober, W. (1997). TGF-beta production regulates the development of the 2,4,6-trinitrophenol-conjugated keyhole limpet hemocyanin-induced colonic inflammation in IL-2-deficient mice. *J. Immunol.* **159,** 3622.

Luo, X., Tarbell, K. V., Yang, H., Pothoven, K., Bailey, S. L., Ding, R., Steinman, R. M., and Suthanthiran, M. (2007). Dendritic cells with TGF-beta1 differentiate naive CD4 + CD25- T cells into islet-protective Foxp3+ regulatory T cells. *Proc. Natl. Acad. Sci. USA* **104,** 2821.

Mackie, S. L., Vital, E. M., Ponchel, F., and Emery, P. (2005). Co-stimulatory blockade as therapy for rheumatoid arthritis. *Curr. Rheumatol. Rep.* **7,** 400.

Maki, T., Gottschalk, R., Wood, M. L., and Monaco, A. P. (1981). Specific unresponsiveness to skin allografts in anti-lymphocyte serum-treated, marrow-injected mice: Participation of donor marrow-derived suppressor T cells. *J. Immunol.* **127,** 1433.

Maki, T., Ichikawa, T., Blanco, R., and Porter, J. (1992). Long-term abrogation of autoimmune diabetes in nonobese diabetic mice by immunotherapy with anti-lymphocyte serum. *Proc. Natl. Acad. Sci. USA* **89,** 3434.

Makino, S., Kunimoto, K., Muraoka, Y., Mizushima, Y., Katagiri, K., and Tochino, Y. (1980). Breeding of a non-obese, diabetic strain of mice. *Exp. Anim.* **29,** 1.

Matsumura, Y., Kobayashi, T., Ichiyama, K., Yoshida, R., Hashimoto, M., Takimoto, T., Tanaka, K., Chinen, T., Shichita, T., Wyss-Coray, T., Sato, K., and Yoshimura, A. (2007). Selective expansion of foxp3-positive regulatory T cells and immunosuppression by suppressors of cytokine signaling 3-deficient dendritic cells. *J. Immunol.* **179,** 2170.

Mellor, A. L., and Munn, D. H. (2008). Creating immune privilege: Active local suppression that benefits friends, but protects foes. *Nat. Rev. Immunol.* **8,** 74.

Nicolls, M. R., Aversa, G. G., Pearce, N. W., Pöspinelli, A., Berger, M. F., Gurley, K. E., and Hall, B. M. (1993). Induction of long-term specific tolerance to allografts in rats by therapy with an anti-CD3-like monoclonal antibody. *Transplantation* **55,** 459.

Ochi, H., Abraham, M., Ishikawa, H., Frenkel, D., Yang, K., Basso, A. S., Wu, H., Chen, M. L., Gandhi, R., Miller, A., Maron, R., and Weiner, H. L. (2006). Oral CD3-specific antibody suppresses autoimmune encephalomyelitis by inducing CD4(+)CD25(-)LAP(+) T cells. *Nat. Med.* **12,** 627.

Ortho Multicenter Transplant Study Group (1985). A randomized clinical trial of OKT3 monoclonal antibody for acute rejection of cadaveric renal transplants. *N. Engl. J. Med.* **313**, 337.

Parlevliet, K. J., Ten Berge, I. J., Yong, S. L., Surachno, J., Wilmink, J. M., and Schellekens, P. T. (1994). In vivo effects of IgA and IgG2a anti-CD3 isotype switch variants. *J. Clin. Invest.* **93**, 2519.

Pearson, T. C., Madsen, J. C., Larsen, C. P., Morris, P. J., and Wood, K. J. (1992). Induction of transplantation tolerance in adults using donor antigen and anti-CD4 monoclonal antibody. *Transplantation* **54**, 475.

Perruche, S., Zhang, P., Liu, Y., Saas, P., Bluestone, J. A., and Chen, W. (2008). CD3-specific antibody-induced immune tolerance involves transforming growth factor-beta from phagocytes digesting apoptotic T cells. *Nat. Med.* **14**, 528.

Plain, K. M., Chen, J., Merten, S., He, X. Y., and Hall, B. M. (1999). Induction of specific tolerance to allografts in rats by therapy with non-mitogenic, non-depleting anti-CD3 monoclonal antibody: Association with TH2 cytokines not anergy. *Transplantation* **67**, 605.

Qin, S., Cobbold, S. P., Pope, H., Elliott, J., Kioussis, D., Davies, J., and Waldmann, H. (1993). "Infectious" transplantation tolerance. *Science* **259**, 974.

Rutella, S., Danese, S., and Leone, G. (2006). Tolerogenic dendritic cells: Cytokine modulation comes of age. *Blood* **108**, 1435.

Salomon, B., Lenschow, D. J., Rhee, L., Ashourian, N., Singh, B., Sharpe, A., and Bluestone, J. A. (2000). B7/CD28 costimulation is essential for the homeostasis of the CD4 + CD25 + immunoregulatory T cells that control autoimmune diabetes. *Immunity* **12**, 431.

Serreze, D. V., and Silveira, P. A. (2003). The role of B lymphocytes as key antigen-presenting cells in the development of T cell-mediated autoimmune type 1 diabetes. *Curr. Dir. Autoimmun.* **6**, 212.

Serreze, D. V., Chapman, H. D., Varnum, D. S., Hanson, M. S., Reifsnyder, P. C., Richard, S. D., Fleming, S. A., Leiter, E. H., and Shultz, L. D. (1996). B lymphocytes are essential for the initiation of T cell- mediated autoimmune diabetes: Analysis of a new "speed congenic" stock of NOD.Ig mu(null) mice. *J. Exp. Med.* **184**, 2049.

Sherry, N. A., Chen, W., Kushner, J. A., Glandt, M., Tang, Q., Tsai, S., Santamaria, P., Bluestone, J. A., Brillantes, A. M., and Herold, K. C. (2007). Exendin-4 improves reversal of diabetes in NOD mice treated with anti-CD3 monoclonal antibody by enhancing recovery of beta-cells. *Endocrinology* **148**, 5136.

Shizuru, J. A., Gregory, A. K., Chao, C. T., and Fathman, C. G. (1987). Islet allograft survival after a single course of treatment of recipient with antibody to L3T4. *Science* **237**, 278.

Smith, J. A., Tso, J. Y., Clark, M. R., Cole, M. S., and Bluestone, J. A. (1997). Nonmitogenic anti-CD3 monoclonal antibodies deliver a partial T cell receptor signal and induce clonal anergy. *J. Exp. Med.* **185**, 1413.

Smith, J. A., Tang, Q., and Bluestone, J. A. (1998). Partial TCR signals delivered by FcR-nonbinding anti-CD3 monoclonal antibodies differentially regulate individual Th subsets. *J. Immunol.* **160**, 4841.

Sreenan, S., Pick, A. J., Levisetti, M., Baldwin, A. C., Pugh, W., and Polonsky, K. S. (1999). Increased beta-cell proliferation and reduced mass before diabetes onset in the nonobese diabetic mouse. *Diabetes* **48**, 989.

Steinman, R. M., Hawiger, D., and Nussenzweig, M. C. (2003). Tolerogenic dendritic cells. *Annu. Rev. Immunol.* **21**, 685.

Strandell, E., Eizirik, D. L., and Sandler, S. (1990). Reversal of beta-cell suppression *in vitro* in pancreatic islets isolated from nonobese diabetic mice during the phase preceding insulin-dependent diabetes mellitus. *J. Clin. Invest.* **85**, 1944.

Takahashi, K., Reynolds, M., Ogawa, N., Longo, D. L., and Burdick, J. (2004). Augmentation of T-cell apoptosis by immunosuppressive agents. *Clin. Transplant.* **18**(Suppl. 12), 72.

Utset, T. O., Auger, J. A., Peace, D., Zivin, R. A., Xu, D., Jolliffe, L., Alegre, M. L., Bluestone, J. A., and Clark, M. R. (2002). Modified anti-CD3 therapy in psoriatic arthritis: A phase I/II clinical trial. *J. Rheumatol.* **29,** 1907.

Van Lier, R. A., Boot, J. H., De Groot, E. R., and Aarden, L. A. (1987). Induction of T cell proliferation with anti-CD3 switch-variant monoclonal antibodies: Effects of heavy chain isotype in monocyte-dependent systems. *Eur. J. Immunol.* **17,** 1599.

Van Wauwe, J. P., De Mey, J. R., and Goossens, J. G. (1980). OKT3: A monoclonal anti-human T lymphocyte antibody with potent mitogenic properties. *J. Immunol.* **124,** 2708.

Wajchenberg, B. L. (2007). Beta-cell failure in diabetes and preservation by clinical treatment. *Endocr. Rev.* **28,** 187.

Waldmann, H., Chen, T. C., Graca, L., Adams, E., Daley, S., Cobbold, S., and Fairchild, P. J. (2006). Regulatory T cells in transplantation. *Semin. Immunol.* **18,** 111.

Wekerle, T., Kurtz, J., Bigenzahn, S., Takeuchi, Y., and Sykes, M. (2002). Mechanisms of transplant tolerance induction using costimulatory blockade. *Curr. Opin. Immunol.* **14,** 592.

Wells, A. D., Li, X. C., Li, Y., Walsh, M. C., Zheng, X. X., Wu, Z., Nunez, G., Tang, A., Sayegh, M., Hancock, W. W., Strom, T. B., and Turka, L. A. (1999). Requirement for T-cell apoptosis in the induction of peripheral transplantation tolerance. *Nat. Med.* **5,** 1303.

Wesselborg, S., Janssen, O., and Kabelitz, D. (1993). Induction of activation-driven death (apoptosis) in activated but not resting peripheral blood T cells. *J. Immunol.* **150,** 4338.

Wofsy, D., Mayes, D. C., Woodcock, J., and Seaman, W. E. (1985). Inhibition of humoral immunity *in vivo* by monoclonal antibody to L3T4: Studies with soluble antigens in intact mice. *J. Immunol.* **135,** 1698.

Wong, J. T., and Colvin, R. B. (1991). Selective reduction and proliferation of the CD4+ and CD8+ T cell subsets with bispecific monoclonal antibodies: Evidence for inter-T cell-mediated cytolysis. *Clin. Immunol. Immunopathol.* **58,** 236.

Wood, M. L., Monaco, A. P., Gozzo, J. J., and Liegeois, A. (1971). Use of homozygous allogeneic bone marrow for induction of tolerance with antilymphocyte serum: Dose and timing. *Transplant. Proc.* **3,** 676.

Woodle, E. S., Xu, D., Zivin, R. A., Auger, J., Charette, J., O'laughlin, R., Peace, D., Jollife, L. K., Haverty, T., Bluestone, J. A., and Thistlethwaite, J. R., Jr. (1999). Phase I trial of a humanized, Fc receptor nonbinding OKT3 antibody, huOKT3gamma1(Ala-Ala) in the treatment of acute renal allograft rejection. *Transplantation* **68,** 608.

Woodside, K. J., Hu, M., Liu, Y., Song, W., Hunter, G. C., and Daller, J. A. (2006). Apoptosis of allospecifically activated human helper T cells is blocked by calcineurin inhibition. *Transpl. Immunol.* **15,** 229.

Xu, D., Alegre, M. L., Varga, S. S., Rothermel, A. L., Collins, A. M., Pulito, V. L., Hanna, L. S., Dolan, K. P., Parren, P. W., Bluestone, J. A., Jolliffe, L. K., and Zivin, R. A. (2000). In vitro characterization of five humanized OKT3 effector function variant antibodies. *Cell Immunol.* **200,** 16.

Yamazaki, S., Bonito, A. J., Spisek, R., Dhodapkar, M., Inaba, K., and Steinman, R. M. (2007). Dendritic cells are specialized accessory cells along with TGF- for the differentiation of Foxp3+ CD4+ regulatory T cells from peripheral Foxp3 precursors. *Blood* **110,** 4293.

Yasunami, R., and Bach, J. F. (1988). Anti-suppressor effect of cyclophosphamide on the development of spontaneous diabetes in NOD mice. *Eur. J. Immunol.* **18,** 481.

Yazdanbakhsh, K., Choi, J. W., Li, Y., Lau, L. F., and Choi, Y. (1995). Cyclosporin A blocks apoptosis by inhibiting the DNA binding activity of the transcription factor Nur77. *Proc. Natl. Acad. Sci. USA* **92,** 437.

You, S., Thieblemont, N., Alyanakian, M. A., Bach, J. F., and Chatenoud, L. (2006). Transforming growth factor-beta and T-cell-mediated immunoregulation in the control of autoimmune diabetes. *Immunol. Rev.* **212,** 185.

You, S., Leforban, B., Garcia, C., Bach, J. F., Bluestone, J. A., and Chatenoud, L. (2007). Adaptive TGF-{beta}-dependent regulatory T cells control autoimmune diabetes and are a privileged target of anti-CD3 antibody treatment. *Proc. Natl. Acad. Sci. USA* **104,** 6335.

Zeiser, R., Nguyen, V. H., Beilhack, A., Buess, M., Schulz, S., Baker, J., Contag, C. H., and Negrin, R. S. (2006). Inhibition of CD4+CD25+ regulatory T-cell function by calcineurin-dependent interleukin-2 production. *Blood* **108,** 390.

CHAPTER 3

GAD65 Autoimmunity—Clinical Studies

Raivo Uibo* and **Åke Lernmark**[†]

Contents		
	1. Introduction	40
	1.1. From 64K to GAD65 (1981–1991)	41
	1.2. GAD65 and GAD67 structure and function	43
	2. GAD65 Autoimmunity in T1D	48
	2.1. Cellular GAD65 immunity in T1D	48
	2.2. Humoral GAD65 immunity in T1D	56
	3. Pathogenesis of GAD65 Autoimmunity	61
	4. Preclinical GAD65 Studies	64
	5. Phase I GAD65 Clinical Studies	64
	6. Phase II GAD65 Clinical Studies	65
	7. Ongoing and Future Clinical Studies	67
	8. Concluding Remarks	68
	Acknowledgments	68
	References	68

Abstract Type 1 diabetes (T1D) in children and particularly in teenagers and adults is strongly associated with autoreactivity to the Mr 65,000 isoform of glutamic acid decarboxylase (GAD65). Autoantibodies to GAD65 are common at the time of clinical diagnosis and may be present for years prior to the onset of hyperglycemia. GAD65 autoantibodies predict conversion to insulin dependence when present in patients classified with type 2 diabetes nowadays more often referred to as patients with latent autoimmune diabetes in the adult (LADA) or type 1,5 diabetes. Analyses of T cells with HLA

---

* Department of Immunology, IGMP, Centre of Molecular and Clinical Medicine, University of Tartu, Tartu, Estonia
[†] Lund University/CRC, Department of Clinical Sciences, University Hospital MAS, 20502 Malmö, Sweden

---

Advances in Immunology, Volume 100 © 2008 Elsevier Inc.
ISSN 0065-2776, DOI: 10.1016/S0065-2776(08)00803-1 All rights reserved.

DRB1*0401-tetramers with GAD65-specific peptides as well as of anti-idiotypic GAD65 autoantibodies suggest that GAD65 autoreactivity is common. The immunological balance is disturbed and the appearance of GAD65 autoantibodies represents markers of autoreactive loss of pancreatic beta cells. Extensive experimental animal research, in particular of the Non-obese diabetic (NOD) mouse, showed that GAD65 therapies reduce insulitis and prevent spontaneous diabetes. Recombinant human GAD65 produced by current Good Manufacturing Practice (cGMP) and formulated with alum was found to be safe in Phase I and II placebo-controlled, double-blind, randomized clinical trials. The approach to modulate GAD65 autoreactivity with subcutaneous immunotherapy (SCIT) showed promise as alum-formulated GAD65 induced a dose-dependent reduction in the disappearance rate of endogenous residual C-peptide production. Additional controlled clinical trials are needed to uncover the mechanisms by which subcutaneous injections of recombinant human GAD65 may alter GAD65 autoreactivity.

1. INTRODUCTION

Autoreactivity to the Mr 65,000 isoform of glutamic acid decarboxylase (GAD65) is tightly associated with type 1 diabetes (T1D). The incidence rate of T1D is increasing worldwide and autoreactivity to GAD65 is ubiquitous, that is, GAD65 autoantibodies (GAD65A) are detectable in T1D patients regardless of whether they are diagnosed in high- or low-incidence countries. GAD65A together with autoantibodies against insulin (IAA) and IA-2 (IA-2A) may account for up to 90% of autoantibody positivity in high-incidence countries. The frequency of GAD65A in newly diagnosed T1D patients is, however, age-dependent (Graham et al., 2002). The older the patient, the more frequent are GAD65A while IAA and IA-2A are more common among infants and children. While IAA and IA-2A would seem to better predict T1D in childhood, GAD65A is a better predictor in adults, in particular of patients classified with latent autoimmune diabetes in the adult (LADA) (for reviews, see Palmer et al., 2005; Wasserfall and Atkinson, 2006). Ever since the discovery of the 64K protein (Baekkeskov et al., 1982), the demonstration that the 64K protein had GAD activity (Baekkeskov et al., 1990), and that human islet GAD was a novel isoform, GAD65 of GAD (Karlsen et al., 1991), it has been of interest to use this autoantigen to test the hypothesis that GAD65 exposure may induce immunological tolerance in subjects with GAD65 autoreactivity. Inducing tolerance to GAD65 might stifle the autoimmune process against the pancreatic islet beta cells. Two studies in the spontaneous diabetic NOD mouse provided proof of principle that GAD65 exposure to mice at risk for spontaneous autoimmune diabetes may inhibit the disease process (Kaufman et al., 1993; Tisch et al., 1993). These studies were soon to be followed by numerous animal investigations including genetic manipulations to demonstrate that GAD65 exposure may inhibit the islet inflammatory process (for reviews, see Bresson and von Herrath, 2007; Rabinovitch, 1994). These preclinical studies have been followed by Phase I and Phase II clinical trials with alum-formulated recombinant human GAD65 to explore the safety of a treatment regimen that would involve a prime and a boost injection with different dosages of alum-formulated GAD65 (Lernmark and Agardh, 2005). In this chapter, we have summarized some of the many experimental and clinical observational studies, which eventually resulted in specific immunotherapy (SIT) clinical trials with alum-formulated GAD65. These trials have been designed to explore the possibility that alum-GAD65 treatment may preserve residual beta-cell function. This is a brief review of the history and immunobiology of GAD65 in T1D with emphasis of the use and safety of recombinant human GAD65 in clinical trials.

1.1. From 64K to GAD65 (1981–1991)

The first description of islet autoantibodies was with classic indirect immunofluorescence using frozen sections of human pancreas from blood group O donors (Bottazzo et al., 1974). These islet cell antibodies (ICA) reacted with the cytoplasmic compartment of all islet endocrine cells. The first type of patients reported was adults with T1D and coexisting autoimmune disease (Bottazzo et al., 1974; MacCuish et al., 1974). Later, ICA was also reported at a high frequency in newly diagnosed T1D children (Lendrum et al., 1976). The pattern of staining revealed beta- (Gianani et al., 1992) or alpha-cell (Bottazzo and Lendrum, 1976)-specific patient sera. The beta-cell-specific pattern was later shown to be GAD65A specific (Marshall et al., 1994) while the autoantigen(s) of the alpha cell staining remains unresolved. It was shown that sera from new onset T1D patients had autoantibodies reacting with living pancreatic islet tumor cells (Maclaren, Huang, and Fogh, 1975) as well as with normal rat islet cells (Huen et al., 1983; Lernmark et al., 1978). These islet cell surface antibodies (ICSA) were also able to mediate complement-dependent cytotoxicity (Dobersen et al., 1980; Radillo et al., 1996). Initially, much research was focused on the detection of autoantibodies while little attention was paid to the actual autoantigen(s) (Lernmark and Baekkeskov, 1981). Using metabolic labeling of isolated human islets (Baekkeskov et al., 1981), it was first demonstrated that sera from new onset T1D patients immunoprecipitated an Mr 64K protein (Baekkeskov et al., 1982) (Table 3.1). In some sera a 38K islet protein was also precipitated. Autoantibodies to the 64K protein were

TABLE 3.1 GAD65 from bench to bedside

1981/1982	Type 1 diabetes (T1D) sera immunoprecipitate a 64K protein	Baekkeskov et al. (1990)
1984	64K antibodies precede clinical onset	Baekkeskov et al. (1984)
1987	64K antibodies predict T1D in first-degree relatives	Baekkeskov et al. (1987)
1990	First confirmation—64K was thought to be an artifact by some authors	Atkinson et al. (1990)
1990	Breakthrough 1: the 64K immunoprecipitate has GAD activity	Baekkeskov et al. (1990)
1991	Breakthrough 2: human islet GAD is GAD65	Karlsen et al. (1991)
1994	GAD65 autoantibodies are detected in a radiobinding assay using GAD65 cDNA to label the protein by *in vitro* transcription translation	Grubin et al. (1994)
1998	Instant confirmation	Schmidli et al. (1994); Schmidli et al. (1995); Verge et al. (1998)
2003	What is the epitope seen by GAD65 autoantibodies?	Padoa et al. (2003)
2005	The middle epitope GAD65 autoantibodies predict T1D	Schlosser et al. (2005)
2007	The GAD65 structure is solved—finally	Fenalti et al. (2007b)
2008	GAD65 anti-idiotypic antibodies are low in T1D	Oak et al. (2008)

found to predict T1D (Table 3.1) and were used as a marker in an immunomodulating clinical trial (Sundkvist et al., 1994).

The 64K protein was eventually reported to have GAD activity (Baekkeskov et al., 1990), which made it possible to discover that human islets expressed a hitherto unknown isoform of GAD65 coded for on chromosome 10 (Karlsen et al., 1991). The previously known GAD was GAD67 coded for on chromosome 2 (Brilliant et al., 1990) and cloning both human GAD67 and GAD65 from human brain mRNA confirmed that the two isoforms were coded by two different genes (Bu et al., 1992). The cloning of GAD65 cDNA made it possible to label the protein by *in vitro* transcription translation (Grubin et al., 1994; Petersen et al., 1994a) and to develop internationally standardized autoantibody assays using a WHO Standard (Mire-Sluis *et al.*, 2000). A standardized GAD65A has been a prerequisite for subsequent SIT clinical trials with recombinant GAD65. Standardized GAD65A assays are also used in prospective natural history clinical trials such as the TEDDY (The Environmental Determinants of Diabetes in the Young) study (Hagopian *et al.*, 2006; TEDDY, 2007). Finally, the expression of human GAD65 cDNA in different systems and purification of the GAD65 protein by current Good Manufacturing Practice (cGMP) have made it possible to begin SIT clinical trials with alum-formulated GAD65 (Agardh *et al.*, 2005; Lernmark and Agardh, 2005).

1.2. GAD65 and GAD67 structure and function

1.2.1. The GAD65 gene

The structural arrangements of the GAD65 and GAD67 genes, *GAD2* and *GAD1*, respectively (Fig. 3.1), have been studied in detail (Bu *et al.*, 1992). The GAD65 cDNA was first cloned from human islets and the gene

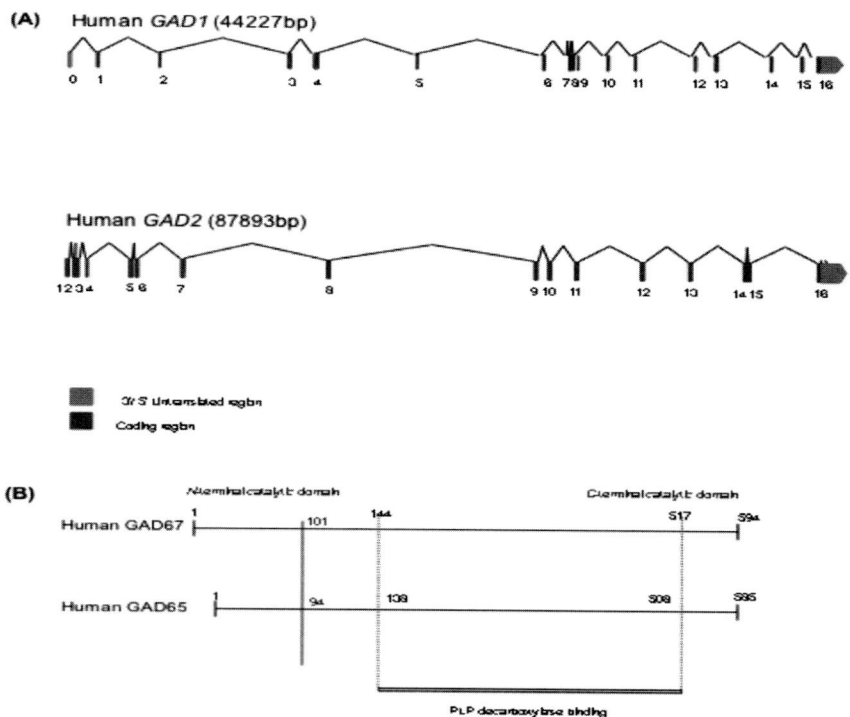

FIGURE 3.1 Genomic structure of GAD1 and GAD2.

located to chromosome 10 (Karlsen et al., 1991). The human GAD65 cDNA encodes a Mr 65,000 polypeptide, with 585 amino acid residues. The coding region of the GAD65 gene, *GAD2*, consists of 16 exons, spanning more than 79 kb of genomic DNA (Bu et al., 1992). Exon 1 contains the 5' untranslated region of GAD65 mRNA, and exon 16 specifies the protein's C-terminal and at least part of the mRNA's 3' untranslated sequence. The GAD67 cDNA had been identified previously and the gene, *GAD1*, located to chromosome 2 (Brilliant et al., 1990). The human GAD67 cDNA encodes a Mr 67,000 polypeptide, with 594 predicted amino acid residues (Bu et al., 1992). The coding region of the GAD67 gene consists of 16 exons in more than 45 kb of genomic DNA. The GAD67 gene contains an additional exon (exon 0) that, together with part of exon 1, specifies the 5' untranslated region of GAD67 mRNA. Exon 16 specifies the entire 3' untranslated region of GAD67 mRNA. The understanding of the cDNA and genetic structure was a prerequisite for the development of methods that made large-scale production of recombinant human GAD65 as well as GAD67. For use in humans it is required that the GAD65 is produced according to cGMP. The cDNA structure allows the expression of full-length GAD65 as the protein apparently is made in free ribosomes without a signal peptide that would sort the protein into the endoplasmic reticulum.

The production of GAD65 is complicated by the N-terminal end sequence which contains a sequence that allows GAD65 to be bound to intracellular membranes as opposed to GAD67 that remains soluble (Dirkx et al., 1995; Solimena et al., 1994). The primary difference between the two isoforms resides in the N-terminal part of the molecule including a GAD65 membrane-anchoring domain. Mutants of the membrane targeting domain spanning amino acids 24–31 of GAD65 affected neither the enzymatic activity nor autoantibody recognition (Plesner et al., 2001). Substituting the membrane targeting domain of amino acids 24–31 may represent an approach to express GAD65 at higher levels for SIT trials. Currently, large-scale production of full-length recombinant human GAD65 is complicated by the fact that the protein needs to be extracted with detergent from either insect (Moody et al., 1995) or methylotrophic yeast, *Pichia methanolica* (Raymond et al., 1998) cells. GAD65 truncated at the N-terminal end is readily produced in *Saccharomyces cerevisiae* yeast cells (Law et al., 1998).

1.2.2. GAD65 primary structure

GAD65 is primarily expressed in neurons and in pancreatic beta cells, where it is concentrated in the Golgi complex region and in proximity to GABA-containing vesicles. The *GAD2* gene for GAD65 codes for 585 amino acids (Fig. 3.2). GAD65, but not the GAD67 isoform, which has a more diffuse cytosolic distribution, is palmitoylated within its first

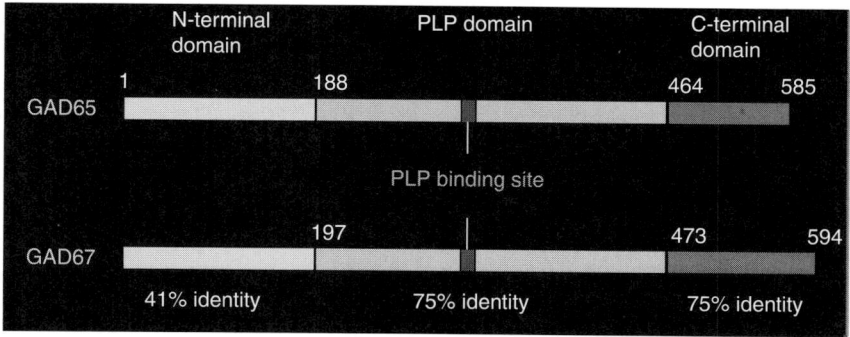

FIGURE 3.2 Primary structure of GAD65 and GAD67. (See Plate 2 in Color Plate Section.)

100 amino acids. The replacement of amino acid residues 1–29 of GAD67 with the corresponding amino acid residues 1–27 of GAD65 was sufficient to target the otherwise soluble GAD67 to the Golgi complex region (Solimena et al., 1994). GAD65 shows about 65% amino acid sequence similarity with GAD67 (Bu et al., 1992; Karlsen et al., 1991). The N-terminal region shows the highest diversity between the two predicted proteins. The deduced pyridoxal-5-phosphate (PLP)-binding site of all GAD molecules is identical throughout evolution. Of the 15 predicted cysteine residues in the *GAD2* gene, 11 are preserved within 13 cysteine residues predicted from *GAD1*. The homology between GAD65 and GAD67 has made it possible to gain a rapid understanding of differences in immunological autoreactivity between the two proteins. GAD65, but not GAD67, is recognized by autoantibodies in T1D (Falorni et al., 1994; Vandewalle et al., 1995) and Stiff Person Syndrome (SPS, formerly called Stiff Man Syndrome) (Bjork et al., 1994). Sera reactive with GAD67 were typically positive for GAD65A. In patients with SPS, the GAD65A titers are often very high and also inhibit the enzymatic activity (Bjork et al., 1994; Raju et al., 2005). SPS sera identified a linear epitope on amino acids residues 4–22 of GAD65 that was recognized solely by autoantibodies from patients with SPS but not by serum from T1D patients (Raju et al., 2005). The autoreactivity in high titer sera with both GAD65 and GAD67 is therefore explained by autoantibodies recognizing sequences or epitopes shared between the two proteins. In developing SIT trials with GAD65 in humans, it is of particular importance to develop specific and highly sensitive means by which to monitor epitope-specific autoreactivity. T-cell reactivity may be monitored with HLA DR tetramer reagents. Rare antigen-specific $CD4^{(+)}$ T cells can now be selectively identified, isolated, and characterized (Mallone and Nepom, 2004). B cell or rather autoantibody reactivity may be monitored with recombinant GAD65 in competition experiments but epitope mapping is most efficiently and reliably carried out with recombinant human GAD65-specific Fab in

competition experiments (Fenalti *et al.*, 2007a; Padoa *et al.*, 2003). Solving of the crystal structures of GAD65 and GAD67 would help to improve our understanding of the mechanisms by which GAD65 autoreactivity develops.

1.2.3. GAD65 crystal structure

The 2.3 Å crystal structure of amino acid residues 84–585 truncated GAD65 was reported using crystals of protein produced in the yeast *Saccharomyces cerevisiae* (Law *et al.*, 1998), which was co-crystallized with chelidonic acid (Fenalti *et al.*, 2007b). GAD65 is an obligate functional dimer and a major surface area is buried in the dimer interface (Fig. 3.3). The co-factor PLP-binding domain has nine alpha-helices, which surround a nine-stranded parallel beta-sheet. PLP is thought to be covalently bound to the catalytic Lys396 residue. In contrast to the highly ordered PLP catalytic domain of GAD67, that of GAD65 is disordered as the active site is fully exposed. This structural difference may explain why GAD65, but not GAD67, is autoinactivated. This means that exposing GAD67 to glutamate does not affect the GAD67 activity while exposing GAD65 results in a loss of activity (Fenalti *et al.*, 2007b). As the C-terminal domain is more mobile in GAD65 compared to GAD67, it was speculated that the C-terminal domain may contribute to the GAD65 autoinactivation (Fenalti *et al.*, 2007b).

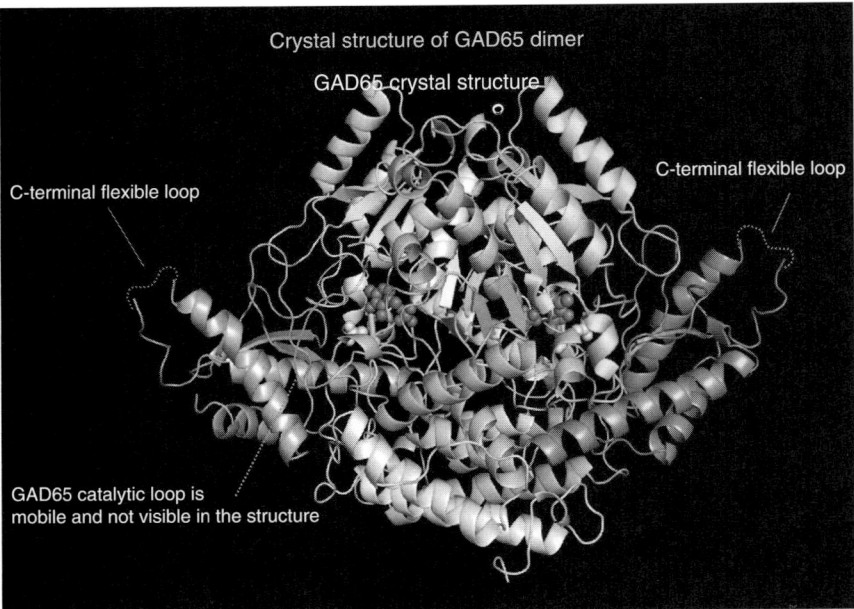

FIGURE 3.3 Crystal structure of GAD65. (See Plate 3 in Color Plate Section.)

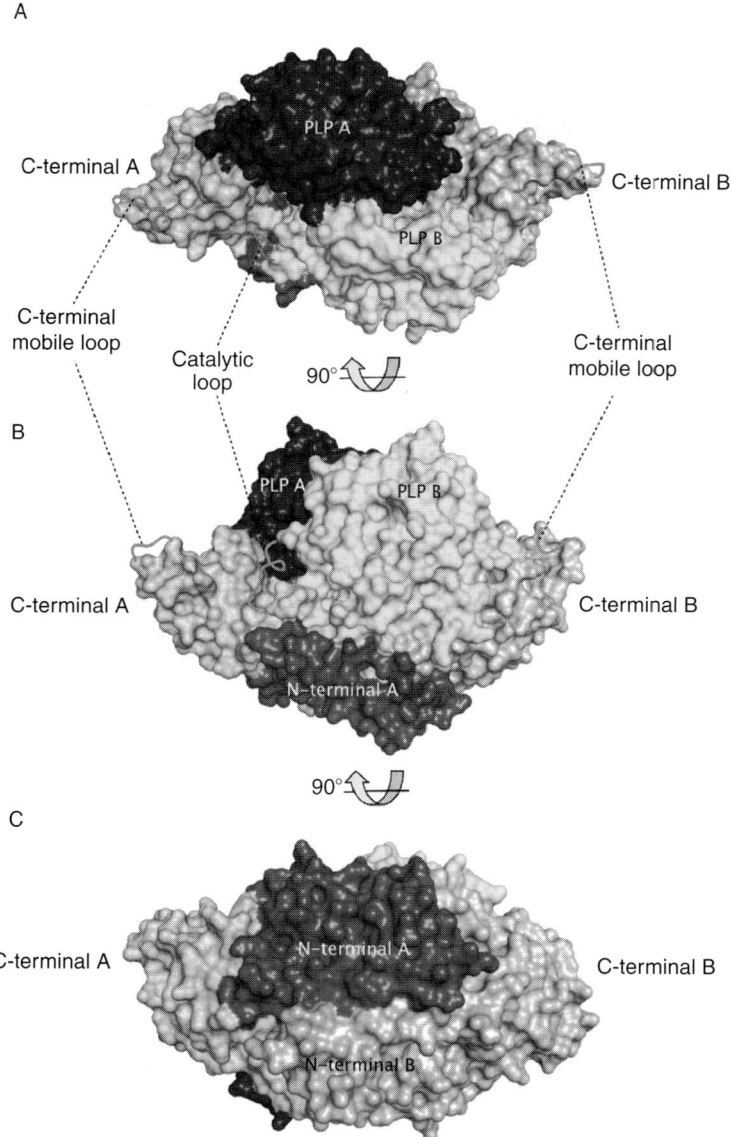

FIGURE 3.4 Domains of GAD65. (See Plate 4 in Color Plate Section.)

Using monoclonal antibodies reactive with the three domains: N-terminal, PLP-binding, and C-terminal (Fig. 3.4), showed that the C-terminal region on GAD65 was the major autoantigenic site. GAD65 autoantibody epitopes were distributed within two separate clusters around different faces of the C-terminal domain. Inclusion of epitope sites in the PLP- and N-terminal domains was attributed to the juxtaposition of all three domains in the crystal structure. T-cell sequences restricted by HLA DRB1*0401 were aligned to GAD65 solvent-exposed regions and colocalized within the two autoantibody epitope clusters. The continuous C-terminal epitope region of GAD65 was structurally highly flexible and therefore differed markedly from the equivalent region of GAD67. The close proximity of B- and T-cell epitopes within the GAD65 structure is important. This observation suggests that antigen–antibody complexes may influence antigen processing by antigen-presenting cells (APC) and thereby T-cell reactivity. This needs to be taken into account in ongoing and future SIT trials with human recombinant GAD65.

2. GAD65 AUTOIMMUNITY IN T1D

After the identification of the 64K protein (Baekkeskov et al., 1982; Lernmark and Baekkeskov, 1981) as GAD65 (Karlsen et al., 1991) and the use of recombinant protein to detect autoantibodies numerous studies have been published on the association between GAD65A and T1D as well as with SPS. While GAD65A determination has found use in the clinic (for a review, see Schmidt et al., 2005), cellular immunoreactivity to GAD65 is less well understood and not yet applicable in the clinic. In this section we will therefore first review aspects of the cellular immunity and GAD65 autoimmunity to be followed by a discussion of the humoral immune responses to GAD65. The emphasis is not necessarily on mechanisms but rather on the importance to understand reactivity to GAD65 in SIT clinical trials in patients with GAD65A.

2.1. Cellular GAD65 immunity in T1D

The lack of reproducible and standardized tests of the cellular immune response to GAD65 has hampered analyses that would improve our understanding of the immune response to injected GAD65. The following is therefore an analysis of existing data, possible approaches, and potential assays to be used in future immunomodulation clinical trials.

2.1.1. Role of antigen presenting cells
In the cascade of events leading to autoimmune damage in T1D, APC play a key role in initiating and regulating immune responses to foreign and self-antigens. Among different APC, dendritic cells (DC) represent a heterogeneous and highly specialized group of cells with various prominent effects on immune response. DC may be divided at least into two groups, conventional and plasmacytoid, distinguished by their reactivity with a panel of CD monoclonal antibodies and function. Convential DC consist of a number of subsets with diverse function and phenotype,

including the most commonly studied myeloid DC. Plasmacytoid DC are characterized mostly by their ability to produce high levels of interferon (INFα) in response to virus.

As DC are responsible to maintain a balance between immunity and tolerance, understanding their role in T1D and in the use of GAD65 for SIT is very important. The crucial question is how tolerance to GAD65 is maintained in healthy individuals but broken in T1D. According to existing understanding, the tolerance to self-antigens is supported by central and peripheral tolerance mechanisms. Central tolerance to self-antigens is obtained by their presentation to maturating T cells in medullary thymus epithelial cells resulting in apoptosis of lymphocyte clones with self-peptide recognizing T Cell Receptor (TCR). The autoimmune regulator AIRE gene has a central but not exclusive role (Gotter and Kyewski, 2004; Pitkanen and Peterson, 2003). The process of central tolerance is not foolproof as some self-reactive T cells escape to the periphery and these cells must be eliminated or kept in control (tolerized) to avoid autoimmunity. Most interestingly, there is no difference in thymic selection of DR4-restricted GAD65 reactive cells among healthy and T1D individuals (Danke et al., 2005).

Much less is known about self-tolerance mechanisms for B cells. In bone marrow, most immature B cells exhibit self-reactivity, whereas in the periphery 20% of mature naïve B cells retain low levels of self-reactivity, predominantly with cytoplasmic antigens. During this maturation process there are three check-points to secure normal B cell self-tolerance and to suppress the development of high-affinity autoantibodies to self-antigens (Shlomchik, 2008; Wardemann and Nussenzweig, 2007).

DC at an immature stage are thought to be tolerogenic and those at mature stage immunogenic for an antigen. In terms of GAD65 little is known about the development of normal tolerance as well as the loss of tolerance in the natural history of T1D. Immunochemical studies suggest that GAD65/GAD67 is expressed in thymus and peripheral lymphoid organs (Pugliese et al., 2001). This expression would seem to contribute to central and peripheral tolerance for these two antigens in healthy persons. However, when GAD65 and GAD67 expression was determined in thymic medullary epithelial cells only GAD67 expression was demonstrated (Gotter and Kyewski, 2004). The mechanisms that break tolerance in T1D are not understood. Recent data suggest that the differentiation and function of DC is abnormal in patients with T1D and their relatives (Jansen et al., 1995; Skarsvik et al., 2004; Takahashi et al., 1998; Vuckovic et al., 2007) perhaps conditioned by abnormal NF-κB function in these cells (Mollah et al., 2008). The consequences of these abnormalities in GAD65 SIT clinical trials remain to be determined.

Self-peptides are typically presented to TCR by HLA class II molecules. Chaperone molecules such as heat-shock proteins may contribute to the process of antigen presentation (Rajagopal et al., 2006). A trimolecular complex is formed between GAD65 peptides and the T1D susceptibility HLA class II molecules. The two major haplotypes associated with T1D, DQ2 (short for DQ A1*0501-B1*0201) and DQ8 (short for DQ A1*0301-B1*0302), may both present GAD65 peptides. As the pattern of presentation of GAD65 peptides by these HLA-DQ heterodimers needs to be clarified it should be noted that GAD65A in both nondiabetic (Rolandsson et al., 2001) and newly diagnosed T1D patients (Graham et al., 2002; Sanjeevi et al., 1995) were associated with DQ2 while IAA and IA-2A were associated with DQ8. Each Major Histocompatibility (MHC) molecule selects for a unique set of peptides in an allele-specific manner, and it has been noticed that GAD65 peptides are readily presented, for example, by DRB1*0401 (Wicker et al., 1996). However, islet endothelium may process and present GAD65 to T cells if cytokines are inducing HLA II class and costimulatory molecules expression (Greening et al., 2003). Cross-presentation by HLA I molecules to T helper cells may also be possible, although this phenomenon remains controversial. It is important to clarify the pathways of antigen presentation when recombinant human GAD65 is injected subcutaneously in SIT clinical trials.

The processing of GAD65 for HLA II molecules by APC was investigated by liquid chromatography–electrospray ionization mass spectrometric analysis to identify HLA DR4 bound peptides in eight patients with recently diagnosed T1D (Nepom et al., 2001). Out of approximately 2,000 DR4-associated peptides, only 3 GAD65 sequences were found and a core epitope representing amino acid residues 555–567 was discovered. Using different methodological approaches (human peripheral blood mononuclear cell, T cell clone, cell-line responses, etc.), several other epitopes in the GAD65 molecule have been detected for CD4 T cells. Only some of these peptides were restricted to HLA-DR3, -DR4, or -DQ8 (see review by DiLorenzo and Serreze, 2005, Fig. 3.5). Much less is known about $CD8^+$ T cell GAD65 epitopes possibly because most $CD8^+$ T-cell responses are short-lived (Martinuzzi et al., 2008) and the GAD65 peptides may be presented by a larger number of Class I molecules coded for by heavy chain genes in linkage disequilibrium with DQ2 and DQ8. GAD65-specific cytotoxic T cells (CTL) recognizing a peptide of amino acid residues 114–123 that binds to HLA-A*0201 were reported in asymptomatic and recent onset T1D patients (Panina-Bordignon et al., 1995). It will be important in clinical trials to determine whether the exposure to alum-formulated GAD65 is able to induce specific $CD4^+$, $CD8^+$, or both cells.

Peripheral APC may take up GAD65 by phagocytosis, macropinocytosis, micropinocyosis, or receptor-mediated endocytosis. GAD65 may be discharged from beta cells into the extracellular space in early phases of diabetes (Waldrop et al., 2007) and thereby be available to APC.

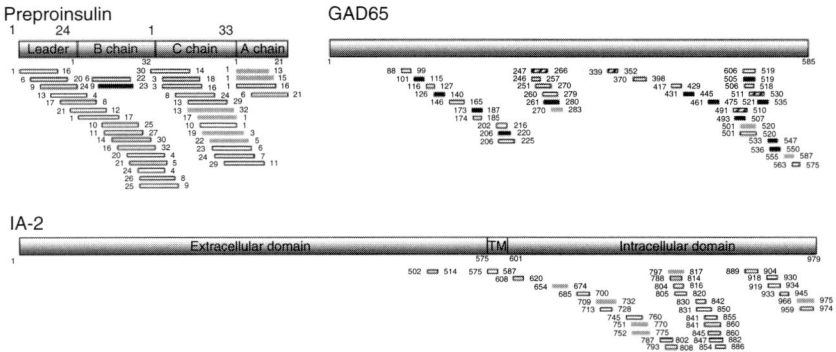

FIGURE 3.5 Representation of GAD65 linear sequence and location of CD4 T-cell epitopes. Similar epitopes are shown for preproinsulin and IA-2 for comparison. Adapted from DiLorenzo et al. (2007). GAD65 and IA-2 are drawn to the same scale, whereas preproinsulin is enlarged threefold for clarity. Where data exist that show an epitope to be unambiguously restricted by a particular HLA class II molecule, this is shown with shading; black is HLA-DQ8; diagonal stripes are HLA-DR3 (DRB1*0301), and grey is HLA-DR4 (DRB1*0401). All other epitopes are shown as white boxes. TM is transmembrane.

DRB1*0401-restricted $CD4^+$ T cells are readily demonstrated among healthy individuals (Danke et al., 2005), and it can therefore not be excluded that the loss of peripheral tolerance to GAD65 may be due to other mechanisms such as molecular mimicry with antigens in microorganisms. Human cytomegalovirus (van der Werf et al., 2007), rubella virus (Ou et al., 2000), and enterovirus (Ellis et al., 2005; Jones and Crosby, 1996; Varela-Calvino and Peakman, 2003) might be involved, although the latter association has been questioned (Sarugeri et al., 2001). It is also possible that GAD65 is altered by transcriptional or posttranscriptional modification that would lead to the loss of tolerance (Chessler and Lernmark, 2000; Namchuk et al., 1997) (Matsukawa and Ueno, 2007). A comprehensive understanding about the handling of the GAD65 by APC is still lacking. Substantial individual differences in GAD65 processing even among patients carrying the same susceptibility allele HLA-DR1*0401 (Reijonen et al., 1999) complicates investigation as does the observation that GAD65 complexed to human serum albumin shows variable T cell proliferative responses (Steed et al., 2008).

It is important to stress that B cells also function as APC. Epstein-Barr Virus (EBV)-transformed DRB1*0401-positive B cells incubated with GAD65 in GAD65A-positive patient sera enhanced proliferation of GAD65-specific T-cell hybridomas (Reijonen et al., 2000). As the effect was blocked by FcR antibodies, it was suggested that processing and presentation of GAD65 peptides was enhanced by GAD65 immune complexes.

On the other hand, GAD65 complexed to monoclonal GAD65 antibodies seemed to block the presentation of the peptide recognized by the monoclonal antibody, thereby generating presentation of other GAD65 epitopes (Jaume et al., 2002). The presentation of minute amounts of GAD65 peptides to CD4$^+$ T helper lymphocytes may be of particular importance to maintain autoimmunity and the autoinflammatory process when most of the target tissue is destroyed (Atkinson et al., 1992; Jaume et al., 2002). These observations complicate understanding the immune response to human recombinant GAD65 used in GAD65A-positive patients in GAD65 SIT clinical trials.

2.1.2. T-cell proliferation studies

To test cellular immunity to GAD65 in T1D, peripheral lymphocyte proliferation studies have been used for more than decade (Atkinson et al., 1992). A number of reports on T-cell responses to GAD65 have been published so far with variable results. The main problem in these investigations is the low responsiveness due to low frequency of GAD65-specific cells in the peripheral blood and the inability of classical T-cell proliferation assays to detect the *in vivo* primed T cells *in vitro*. Also, the results depend much on GAD65 purity, serum used in studies, and method of detection. Therefore, measuring the proliferation of GAD65-specific T cells has remained a significant technical challenge. Accordingly, it is understandable why some earlier studies showed weak and inconsistent results. Also, it is evident that naïve T cells may be effectively primed *in vitro*, which makes the discrimination between naïve and *in vivo* GAD65 primed memory cell responses by simple proliferation assays problematic (Ott et al., 2004). Therefore, traditional *in vitro* proliferation assays have generally not detected marked differences in GAD65-specific T cells when T1D patients and healthy controls were compared (Roep et al., 1999). However, improvements in GAD65 quality and culture conditions improved reproducibility (Tree et al., 2004). If T-cell proliferation in response to GAD65 was measured by flow cytometry-based dilution of the fluorescent dye 5,6-carboxylfluoresein diacetate succidimyl ester (CFSE dilution), there was, however, no difference between T1D and healthy controls (Monti et al., 2007). The latter assays permit further characterization of proliferating cells by their surface phenotype, that is, to separate naïve and memory T-cell responses *in vitro*. It was concluded that patients with T1D and at-risk subjects but not control subjects have GAD65 reactive T cells that are fully autoantigen experienced (Monti et al., 2007). In future studies of peripheral blood T cells obtained in GAD65 SIT clinical trials it will be required that tests show little inter-laboratory variability and exclude naive T-cell activation *in vitro*.

2.1.3. Phenotyping antigen-specific T cells

The procedures for antigen-specific T-cell phenotyping and separation are complicated by the fact that peripheral blood self-antigen reactive T cells are very rare (1 in 10,000–1,000,000) and often seem to have TCR of low avidity to self-antigen (Mallone and Nepom, 2005). Nevertheless, the development of multimeric peptide–MHC complexes technology to detect and analyze the antigen-specific T lymphocytes *in vitro* has increased the possibility to phenotype GAD65-specific $CD4^+$ and $CD8^+$ T cells. Tetramers for enumeration, characterization, and isolation of peptide specific T cells have afforded many advantages over previous techniques (Mallone and Nepom, 2005). They enable direct identification of antigen-specific T cells with minimal *in vitro* manipulation. The GAD65 tetramer technique (Fig. 3.6) was first used to characterize T cells with various epitope specificities in T1D patients and controls (Reijonen *et al.*, 2002). So far mostly HLA-DR4-GAD65 tetramers have been explored to demonstrate that the GAD555–567 epitope is a dominant epitope in T1D confirming the results of proliferation studies (Oling *et al.*, 2005; Reijonen *et al.*, 2004). $CD4^+$ and $CD8^+$ GAD65 reactive T cells of memory (CD45RO+) as well as naïve (CD45RO–) phenotype seems to be present in the peripheral blood of both healthy individuals at risk and patients with T1D, whereas healthy controls have only the GAD65 reactive CD45RO– population of T cells (Danke *et al.*, 2005; Monti *et al.*, 2007). It has been also demonstrated that GAD65 reactive T cells from patients with T1D but not from normal subjects are independent of costimulation through the CD28/B7 pathway (Viglietta *et al.*, 2002) and that their avidity to MHC–peptide is high (Mallone and Nepom, 2004; Reijonen *et al.*, 2004).

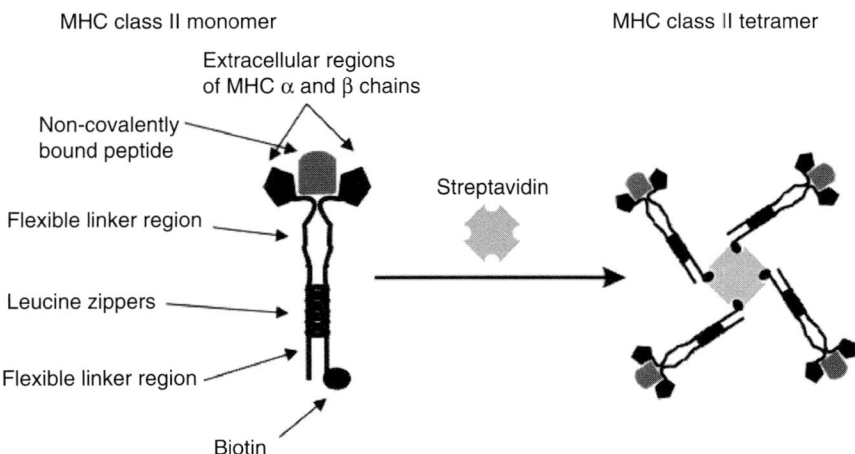

FIGURE 3.6 Structure of MHC Class II tetramers.

Autoantigen-specific T cells that recognized multiple GAD65-derived peptides and coexpressed CD25+ CD134+ were confined to patients and prediabetic subjects, whereas autoantigen-reactive T cells in control subjects were CD25+ CD134– and recognized fewer autoantigen-derived peptides (Endl et al., 1997). CD 134 (Ox40) is an inducible co-receptor molecule thought to be critical for the survival and expansion of inflammation-mediating T cells. These studies underline the fact that GAD65-sensitized peripheral blood T cells have substantially different phenotypes in healthy controls and patients with T1D.

2.1.4. Phenotyping antigen-specific regulatory T cells

It has been demonstrated that T cells of a variety of lineages could suppress antigen-specific cellular responses by a variety of mechanisms including both contact-independent (mediator-dependent) and contact-dependent mechanisms (Bach, 2003). The first mechanism is achieved by regulatory T cells of $CD4^+$ CD25+ FOXP3+ phenotype. There is no significant difference in the frequency of regulatory T cells in T1D patients and control subjects (Brusko et al., 2007; Tree et al., 2004). However, it has become clear that phenotyping of T cells does not reflect their full functional regulatory potential. The suppressor character of these cells could at best be detected by the production of cytokines possessing significant regulatory function *in vivo*, that is, producing IL-10, TGFβ, and their modulators such as IL-21 and others.

The cellular activation to induce cytokine production *in vitro* generally results in down-regulation of T-cell receptors, which puts methodological limitations in revealing antigenic-specific regulatory T cells by tetramer assays. Therefore, traditional methods are used. Thus the depletion of regulatory $CD4^+$ CD25+ T cells from peripheral blood lymphocyte cultures significantly changes the GAD65 stimulatory effect in controls and T1D patients. In controls but not among patients the responses to GAD65 peptide 555–567 increased significantly after regulatory T cells depletion; however, the GAD65 specificity of the depleted cells remained unknown (Danke et al., 2005). Newly diagnosed T1D children had higher mean numbers of IL-10 producing cells in response to DR3-restricted GAD65 epitope 335–352 than controls showing the probable increase in suppressive T cells in the former group (Sanda et al., 2008). Similarly, children born to T1D mothers had possibly been primed with GAD65 during their fetal life and had developed fetal regulatory T cells that might protect them from T1D (Holm et al., 2006). As an example of a contact-dependent mechanism, $CD4^+$ T cell responses to GAD65 in healthy individuals were reported to be regulated by $CD8^+$ suppressor T cells of CD45RA+ CD27– population. Also, they suggested a parallel influence of cytokines produced by $CD8^+$ T cells, thus claiming that the cooperation of multiple subsets of regulatory cells is necessary to maintain self-tolerance (James et al., 2008). These types of assays should prove useful in evaluating effects of GAD65 SIT in clinical trials.

2.1.5. ELISPOT

Enzyme-linked immunospot (ELISPOT) serves as a sensitive tool to detect rare antigen-reactive T-cells phenotype and function *ex vivo* by their ability to secrete cytokines. This assay allows visualization of the secretory product at the single cell level, which makes it informative to analyze antigen-specific cells of different phenotypes. Accordingly, the ELISPOT assay provides both qualitative (type of secreted protein) and quantitative (number of responding cells) information. However, because of high sensitivity, ELISPOT is heavily influenced by methodological conditions. Nonspecific reactions may be reduced by removal of immunoglobulin from the culture medium (Kotani et al., 2002). As a result, significant IFNγ responses to GAD65 peptide of 554–575 amino acids were detected in T1D patients compared to controls (Nagata et al., 2004). From earlier proliferation and tetramer studies (Arif et al., 2004; Reijonen et al., 2002), this GAD65 peptide represent an epitope known to be specific for CD4$^+$ cells in T1D.

The counting of CD8$^+$ T cells by their INFγ production after GAD65 stimulation in ELISPOT is, however, much more difficult due to the rapid disappearance of most antigen-specific INFγ producing CD8$^+$ cells after T1D diagnosis and a possible shift of CD8$^+$ cells toward different target antigens (Martinuzzi et al., 2008). Nevertheless, using ELISPOT two autoantigenic epitopes, GAD65 peptide 114–123 and peptide 536–545, presented by HLA-A2 class alleles (HLA-A*0201) were demonstrated in a substantial proportion of T1D patients but not in control subjects (Blancou et al., 2007).

A cross-sectional study comparing T-cell responses to a DR3-restricted GAD65 peptide 335–352 epitope found an increase in antigen-specific IL-10 producing cells in new-onset patients but a decrease during the T1D honeymoon period (Sanda et al., 2008). This T-cell study is of particular interest since GAD65A are strongly associated with HLA-DR3.

2.1.6. T and B cell interactions

Interactions between T and B cells may be studied through different strategies but studies are dominated by investigating the role of B cells as APC in T1D. Thus, GAD65-specific autoantibodies enhance the presentation of GAD65 to T cells (Reijonen et al., 2000). Using EBV-immortalized B cell lines to GAD65 from HLA-DR3 and DR4 positive new onset T1D patients it was possible to demonstrate B cell presentation of GAD65 peptides to T cells. In this process, GAD65 peptide uptake was likely mediated by B-cell receptor GAD65 IgM (Banga et al., 2004). It is possible that GAD65 uptake by B cells is important for further clustering of T and B cell GAD65 determinants and modulating T-cell responses. Importantly, four of five major T-cell epitopes restricted by HLA-DRB1*0401 were detected in the structure of GAD65 within the same regions as immunodominant B-cell epitopes (Fenalti et al., 2008). In GAD65 SIT clinical trials, the subjects are GAD65 autoantibody positive, which need to be taken into account when peripheral T-cell responses are investigated.

2.1.7. Immunology of diabetes (IDS) T-cell workshops
To overcome the methodological problems and the poor inter-laboratory variation related to the detection of rare autoantigen-reactive T cells in T1D, the IDS T Cell Workshops have been organized (Roep et al., 1999). While blinding and distribution of clinical samples is difficult, workshop organizers generated a set of autoantigenic peptides that represented known or putative epitopes from the major diabetes-related islet autoantigens, including GAD65. During the last 10 years many deficiencies in T-cell assay formats and antigen preparations have been identified, including the demonstration of suitability of immunoglobulin-free ELISPOT to demonstrate the significant IFNγ production in response to GAD65 peptide 554–575 in T1D (Nagata et al., 2004). In 2006, a report on $CD4^+$ $CD25+$ T cells in T1D was published (Tree et al., 2004).

2.2. Humoral GAD65 immunity in T1D
Whatever the autoimmune disease, autoantibodies have always been in the centre of clinical interest due to their potential value in autoimmunity screening, diagnosis as well as monitoring treatment effectiveness and prognosis. In addition, autoantibodies have been effective in the search for autoantigenic molecules and in deciphering unknown pathways. In contrast to the cellular arm of the immune response to GAD65, there is a wealth of studies reporting frequencies of GAD65A in a variety of clinical conditions. The rapid development of GAD65A analyses in the clinic has been spurred by the early effort of assay standardization.

2.2.1. Autoantibody assays (RBA, ELISA, TRF-ELISA)
The molecular characteristics of islet autoantigens remained unknown until it was demonstrated that antibodies in new-onset T1D patients precipitated 35S-methionine-labeled solubilized human pancreatic islet cell proteins of 64K and 38K (Baekkeskov et al., 1982). Later on the 64K protein was identified as GAD65. Several assays have been developed for GAD65A in humans; however, a limited set of tests has remained in use. This is because T1D autoantibodies against GAD65 developed at rather low concentrations against conformational epitopes of the antigen. These features limit GAD65A assay selection.

Currently, three autoantibody assays are most widely used: radiobinding assays (RBA), enzyme-linked immunosorbent assay (ELISA), and time-resolved fluorescence (TRF) ELISA (Fig. 3.7). In addition, some assays take advantage of the simultaneous detection of GAD65 and insulinoma-antigen 2 (IA-2) autoantibodies; however, such assays may have diminished sensitivity and specificity, as shown by recent comparative analysis of GAD65 autoantibody tests (Torn *et al.*, 2008). *In vitro* transcription and translation of recombinant GAD65 with reticulocyte lysate is used to produce 3H or 35S labeled GAD65 in RBA (Grubin *et al.*, 1994; Petersen *et al.*, 1994a). Alternatively, RBA use 125I-labeled GAD65 (Powell *et al.*, 1996). Usually the assays use protein A Sepharose beads to precipitate the antigen–antibody complex to separate bound from free GAD65. Using the *in vitro* transcribed/translated 35S labeled GAD65 in combination with 96-well test plates and effective beta-counters allows tests of several hundreds of serum samples per day.

Classical ELISA for GAD65A has a reduced specificity and sensitivity because GAD65 epitopes required for autoantibody binding are mostly

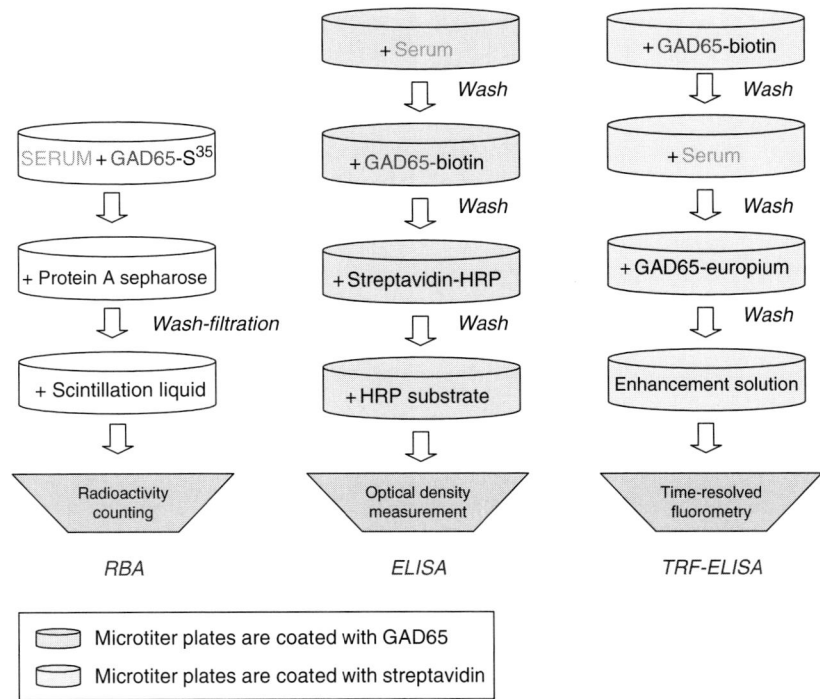

FIGURE 3.7 General principle of radiobinding assay (RBA), enzyme-linked immunoassay (ELISA), and time-resolved fluorescence ELISA (TRF-ELISA) for the detection of antibodies to GAD65. HRP: horseradish peroxidase. (See Plate 5 in Color Plate Section.)

destroyed. For the same reason, GAD65 autoantibodies in T1D sera are also undetectable by immunoblotting. However, recently modified ELISA have overcome these problems (Brooking et al., 2003). The disadvantages of current ELISA kits are that they are generally more costly and require larger serum volumes than in-house RBA using 35S labeled GAD65 (Torn et al., 2008).

Time-resolved fluorescence (TRF) ELISA is less often used to detect GAD65A in spite of some advantages over RBA, such as faster performance and the wider linear range of autoantibody detection. Still TRF-ELISA has not gained usage possibly due to its relatively high cost and less favorable performance in international standardization efforts (Ankelo et al., 2007).

There are also other methods applicable for GAD65 autoantibody determination from T1D sera, such as indirect immunofluorescence on cells transfected with the human GAD65 gene (Papouchado et al., 1996; Villalba et al., 2008), recent attomolar sensitive GAD65 antibody capture assay (Lee et al., 2007), and others. Whether recently proposed liquid phase luminescent immunoprecipitation assay for the detection of GAD65 autoantibodies in SPS (Burbelo et al., 2008) is usable in evaluation of these autoantibodies in T1D needs further study. In SPS, as in type 1 polyendocrinopathy, GAD65 autoantibodies develop against linear rather than conformational epitopes and are demonstrable at much higher dilutions than in T1D (Bjork et al., 1994; Tuomi et al., 1996). There are comprehensive reviews on methods and clinical associations of GAD65A (Pihoker et al., 2005) that are useful when considering GAD65A as an end-point in GAD65 SIT clinical trials.

2.2.2. IDW workshops and DASP

GAD65A assays are increasingly used in clinical and research laboratories either as in-house assays or as commercial kits. However, to provide a reliable marker for T1D, GAD65A must be analyzed in reproducible assays with high diagnostic sensitivity and specificity. In order to overcome the problems related to test standardization, the Diabetes Autoantibody Standardization Program (DASP) was established on the basis of a series of Immunology of Diabetes Society (IDS) workshops and GAD Antibody workshops (Schmidli et al., 1994; Verge et al., 1998). A WHO standard for GAD65A has been established (Bingley et al., 2003; Mire-Sluis et al., 2000). Since the first DASP 2000 (Bingley et al., 2003), follow-up workshops were carried out in 2002, 2003, 2005, and 2007. Results are available on workshops until 2005 (Torn et al., 2008). In general, GAD65A assays achieved acceptable inter-laboratory variation with a high level of sensitivity with good discrimination between health and disease. Problem laboratories or assays are easily identified. Over three DASP workshops (2002–2005) there has been an improvement in the performance of GAD65 autoantibody assays. The combined receiver operator characteristic (ROC) curves demonstrated that, on average, using the threshold 18 WHO units/ml 98% specificity and 88% sensitivity is achieved for newly diagnosed T1D (Torn *et al.*, 2008). Continuous standardization and harmonization of GAD65A is a prerequisite for GAD65 SIT clinical trials.

2.2.3. GAD65 autoantibodies and T1D

As mentioned before, the most commonly used RBA detects autoantibodies by immunoprecipitation using Protein A Sepharose. Protein A effectively binds human IgG subtypes except IgG3. Therefore, RBA with Protein A does not yield information on IgG3, IgM, IgA, and IgE autoantibodies. Instead, subclass-specific monoclonal antibodies in combination with streptavidin-biotin binding to agarose (Hawa *et al.*, 2000; Hoppu *et al.*, 2004; Petersen *et al.*, 1999; Ronkainen *et al.*, 2006) or capture ELISA (Couper *et al.*, 1998) are used. T1D progressors differed from nonprogressors in one study of GAD65A isotypes and IgG subtypes (Ronkainen *et al.*, 2006). The IgG3-type autoantibodies were often characteristic for the first phase of the humoral response to GAD65, while IgG4 antibodies appeared as the last IgG subclass. In another study, nonprogressors had higher levels of IgE and IgM GAD65A than progressors to T1D (Petersen *et al.*, 1999). However, in these studies the sample numbers have been relatively small and blood sampling has not been frequent enough to disclose the possible role between isotype and subtype switch in T1D. It will be of interest to use isotype and subtype switch as secondary outcomes in GAD65 SIT clinical trials.

In addition to GAD65A isotypes and subtypes, the antibody epitopes could give important information not only in T1D pathogenesis but also in response to GAD65 SIT. It has been shown that GAD65A in T1D are predominantly directed to conformational epitopes located in the middle region of GAD65 (Bjork *et al.*, 1994; Hampe *et al.*, 2000; Kobayashi *et al.*, 2003). Also, dynamic changes in GAD65 epitope reactivities during the development of T1D have been demonstrated (Hampe *et al.*, 1999). The epitope of the middle region of GAD65 is recognized in the early stages of GAD65 autoimmunity, while in the later stages the epitopes may spread to the N-terminus of GAD65 (Bekris *et al.*, 2007; Bonifacio *et al.*, 2000; Ronkainen *et al.*, 2006; Schlosser *et al.*, 2005). In the Phase II clinical trial with LADA patients, none of the alum-GAD65 dosages tested seem to affect the GAD65A epitope present at baseline (Bekris *et al.*, 2007).

2.2.4. Human monoclonal GAD65 antibodies

Monoclonal antibodies and recombinant Fab derived from monoclonal antibodies have been particularly informative in studies of GAD65 conformational epitopes (Padoa *et al.*, 2003; Schlosser *et al.*, 2005). During early studies islet-cell reactive human monoclonal antibodies were often IgM and showed cross-reactivity of various human tissues. IgG autoantibody producing B cells, immortalized with EBV, was eventually produced from adults with T1D (Richter et al., 1992). Six monoclonal GAD65 antibodies called Monoclonal Islet Cell Antibodies (MICA) (MICA1, 2, etc.) were produced. Later additional monoclonal GAD65 antibodies were obtained (Madec et al., 1996; Syren et al., 1996; Tremble et al., 1997) and MICAs were further designated as M1, M2, etc. (Schwartz et al., 1999). The availability of recombinant Fabs from GAD65-specific monoclonal antibodies has enlarged our possibilities for epitope mapping and precise identification of disease-specific epitopes (Fenalti et al., 2007a, 2008; Padoa et al., 2003; Raju et al., 2005; Schlosser et al., 2005). It has become evident that two clusters of epitopes, ctc1 and ctc2, are demonstrable in GAD65 (Table 3.2). The former is recognized by monoclonal antibodies b96.11 and appears to be associated with the development of T1D, whereas another binds to b78 monoclonal antibody and to autoantibodies either unassociated with T1D or associated with milder diabetes phenotype (Fenalti et al., 2007a). It should be possible to use these monoclonal GAD65 antibody reagents to monitor responses to human GAD65 treatment in human trials similar to what has been done in analyzing the

TABLE 3.2 Reactivity of 11 monoclonal antibodies with GAD65 epitopes and the C-terminal clusters 1 and 2 (according to Fenalti et al., 2008)

Target area in GAD65	Designation of the antibody	Epitope region	Reactive epitope(s) associate(s) with ...
NH_2-terminus	DPB^a	1–102	–
	DPD^a	96–173	$Ctc1^b$
PLP domain	$b96.11^c$	308–365	$Ctc2^d$
	$M4^e$	308–365	–
	$M6^e$	242–282	–
PLP and NH_2-terminus	DPC^a	134–242, 366–413	Ctc1
COOH-terminus	$M3^e$	483–499, 556–585	Ctc2
	DPA^a	483–499, 556–585	Ctc2
	$M2^e$	514–528	Ctc1
	$M5^e$	512–540	Ctc1
	$b78^c$	532–540, 514–528	Ctc1

[a] Monoclonal antibodies prepared by Madec et al. (1996).
[b] C-terminal cluster 1.
[c] Monoclonal antibodies prepared by Tremble et al. (1997).
[d] C-terminal cluster 2.
[e] Monoclonal antibodies prepared by Richter et al. (1992, 1995).

outcome of the Phase I clinical trial with alum-formulated recombinant human GAD65 (Bekris et al., 2007).

2.2.5. Anti-idiotypic GAD65 antibodies in T1D

Whether antibodies reacting to antigen-binding areas of autoantibodies (anti-idiotype) are involved in pathogenic autoimmunity development has for years been a disputable question in understanding the natural course of autoimmune diseases (Shoenfeld, 2004). The characterization of such antibodies, closely related to the disease pathological mechanisms, and their regulation offer novel tools for analyzing effects of GAD65 SIT and possibly treating T1D. Therefore, the recent demonstration of differences in GAD65 anti-idiotypic antibody incidence between T1D patients and healthy individuals is of particular significance in T1D (Oak et al., 2008). While healthy individuals and first-degree relatives of T1D patients tested GAD65A negative in RBA, they showed GAD65A after the sera were first absorbed with epitope-specific monoclonal GAD65 antibodies, suggesting that GAD65A are masked by epitope-specific anti-idiotypic antibodies. Most important, GAD65A-positive T1D showed a lack of anti-idiotypes to disease-associated monoclonal anti-GAD65 antibody, b96.11. Whether anti-idiotypic immunoglobulins to the b96.11 idiotype may be able to neutralize circulating GAD65A and influence the course of the disease following GAD65 SIT needs further investigations.

3. PATHOGENESIS OF GAD65 AUTOIMMUNITY

The pathogenesis of islet autoimmunity in general as well as GAD65 autoimmunity in particular remains poorly understood. The question always asked is whether autoimmune reactions are of primary or secondary importance for tissue destruction. Based on the fact that (pro)insulin autoantibodies often appear before GAD65A in childhood T1D, it has been suggested that (pro)insulin might be the primary initiating autoantigen in T1D (reviewed in Eisenbarth et al., 2002). Indeed, (pro)insulin is an islet beta cell-specific autoantigen and may therefore explain the specific loss of beta cells in T1D. Similar to (pro)insulin in autoimmune diabetes, thyroid peroxidase in autoimmune thyroiditis (Weetman, 2004) and H+ K+ ATPase in autoimmune gastritis (Toh and Alderuccio, 2004) are specific for the tissues where the autoimmune destruction takes place. GAD65 on the other hand is also present in other cells outside of the pancreatic islets.

Some microbes that commonly induce inflammation in pancreatic tissue have cross-reactivity to GAD65 at the T-cell level and could therefore readily initiate the immune attack against cells that express GAD65. Indeed, several clinical, epidemiological, and immunological investigations have shown viral infections to be associated with T1D development

(Hyoty and Taylor, 2002; van der Werf et al., 2007). However, further systematic long-term analysis of T-cell responses to viral structural components and GAD65 is needed in individuals carrying different T1D susceptibility HLA haplotypes.

It is possible that an overproduction of GAD65 will sensitize lymphocytes and activate an autoimmune process in response to a decrease in local tolerance.

Other environmental factors may also modulate the development of autoimmune reaction development against GAD65. Thus, early cessation of breast-feeding and high consumption of fresh cow's milk has been shown to be associated with GAD65A development in Swedish children (Hummel et al., 2002) (Wahlberg et al., 2006). Gluten (gliadin) does not seem to be involved in GAD65A development (Hummel et al., 2002; Uibo et al., 2001).

Based on these considerations it is suggested that the GAD65-related autoimmunity and tissue destruction in T1D might start by the activation of the T cells by external antigen (Fig. 3.8). In genetically susceptible individuals this leads to a cascade of immunopathogenetic changes eventually resulting in a GAD65-specific lymphocytic infiltration within the pancreatic islets (insulitis). The infiltration of the islets is composed of effector ($CD4^+$ and $CD8^+$ cells, NK cells) and regulatory lymphocytes (induced Tregs and others) originating from pancreatic lymph nodes where T cells are activated by APC presentation of GAD65. This is consistent with the absence of insulitis in GAD65A positive nondiabetic subjects (In't Veld et al., 2007). During islet inflammation accompanied by beta-cell damage and release of cell contents, GAD65 (and other autoantigens) may be taken up by activated endothelial cells that are capable of processing and presenting disease-related epitopes of the GAD65 to autoreactive T cells (Greening et al., 2003).

The characteristics of effector and regulatory as well as other inflammatory cells are determined by the type of mediators produced by these cells and by receptors that are expressed on these cells. The autoantigen-specific $CD4^+$ effector T cell lineages of Th1 subtype producing IFNγ, TNFα, IL-2, IL-12, and IL-18 (Gutcher and Becher, 2007) and Th17 producing pathogenic IL-17 and IL-21 (Nurieva et al., 2007) but possibly also IL-22 may have a central role in initiating and maintaining the beta-cell destruction. The activation and expansion of inflammatory $CD4^+$ cells is promoted by the cytokine milieu generated mostly by the cells of the innate immune system, for example, through IL-6, IL-12, IL-23, and others. In contrast, regulatory T cells producing either IL-10 or TGFβ suppress the function of these inflammatory cells at the site of inflammation or by preventing the priming of naïve autoreactive T cells by APC in secondary lymphoid tissue (Korn and Oukka, 2007). Similarly, a change in the Th1-to-Th2 balance toward IL-4, IL-5, IL-13, and IL-25 producing

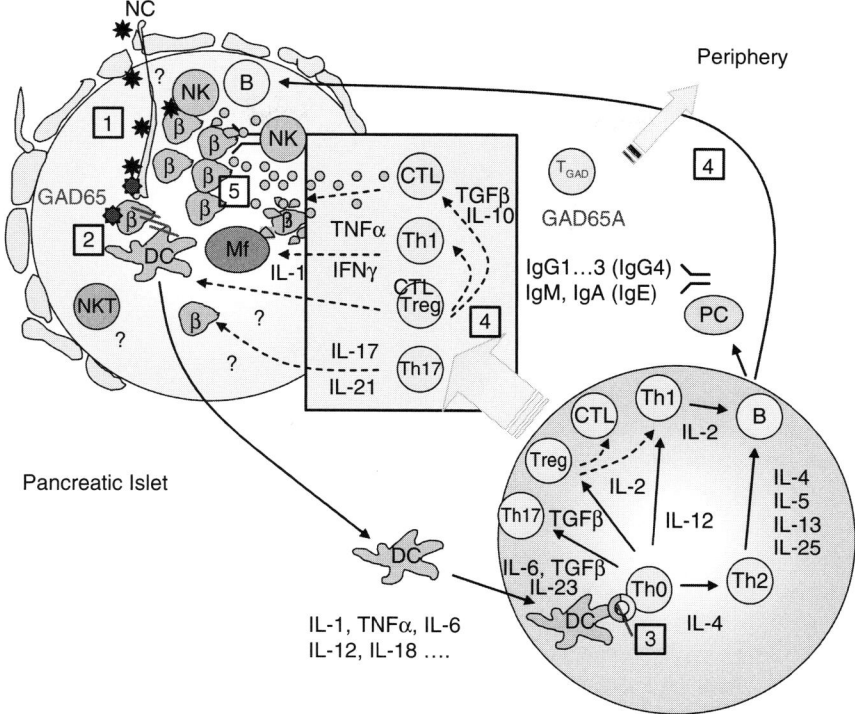

FIGURE 3.8 Schematic view on possible immunopathogenesis of pancreatic islet beta-cells destruction and the development of humoral and cellular immune reactions against GAD65. Steps of events: 1: environmental factors and neural cell (NC) products are conditioning the relevant milieu by activation of dendritic cells (DC), macrophages (Mf), natural killer (NK) cells, natural killer T cells; 2: intake of GAD65 or cross-reactive peptides by DC; 3: presentation of peptides to naive T helper (Th0) cells and subsequent activation and proliferation of type 1 (Th1) and type 2 (Th2) helper cells, IL-17 producing helper cells (Th17), regulatory T (Treg) cells, cytotoxic T cells (CTL), B and plasma cells (PC), activation of different cell subsets by cytokines; 4: migration of activated cells from pancreatic lymph node into pancreatic islet, cross-talk with periphery; 5: destruction of pancreatic beta-cells by cytokine- and perforine/granzyme mediated mechanisms environmental factor (virus, etc.) GAD65, GAD65 or cross-reactive peptide. T cell carrying receptor for GAD65 peptide(s). Antibodies to GAD65. (See Plate 6 in Color Plate Section.)

Th2 cells decreases the inflammatory activity in the pancreatic islet tissue. Nevertheless, no evidence about GAD65 or other islet autoantigen-specific cells phenotypes and function is available from the pancreatic islets of T1D patients. In patients with T1D, infiltration of $CD4^+$, $CD8^+$ T cells, B cells, and macrophages was observed in pancreatic biopsy specimens (Itoh et al., 1993; Uno et al., 2007).

4. PRECLINICAL GAD65 STUDIES

The NOD mouse and the BB rat develop insulin-dependent diabetes following spontaneous appearance of insulitis that precedes clinical onset of hyperglycemia. GAD65 treatment in the BB rat deviated thymic T-cell cytokine expression (Bieg *et al.*, 1997) but did not affect development of T1D (Petersen *et al.*, 1994b). In contrast, NOD diabetes is associated with GAD65 reactive T cells (Petersen *et al.*, 1994b) and GAD65 or GAD65 peptide administration effectively prevented autoimmune destruction of beta cells (Kaufman *et al.*, 1993; Sai *et al.*, 1996; Tisch *et al.*, 1993). Administration of GAD65 by injection (Kaufman *et al.*, 1993) or by nasal administration (Tian *et al.*, 1996a) also inhibited disease progression and prolonged islet graft survival in the NOD mouse (Tian *et al.*, 1996b). Expression of anti-sense GAD65 prevented diabetes in transgenic NOD mice and insulitis was induced by GAD65 peptide-specific and HLA-DQ8-restricted $CD4^+$ T cells from human DR-DQ transgenic mice (Abraham *et al.*, 2000). Numerous studies demonstrate that (1) GAD65 is a key target antigen in the induction of NOD mouse T1D, (2) autoimmunity to GAD65 triggers T-cell responses to other beta-cell antigens, and (3) spontaneous autoimmune disease may be prevented by tolerization to GAD65. Numerous studies have replicated the basic finding of a delayed onset (for a review, see Bresson and von Herrath, 2007) and an international NOD mouse workshop was carried out to test interlaboratory and NOD mouse line variation in GAD65 autoreactive T-cell responses (Kaufman *et al.*, 2001). It is important that the many molecular genetics studies in the NOD mouse are far from translation to human studies because of safety.

5. PHASE I GAD65 CLINICAL STUDIES

Diamyd Medical AB Stockholm, Sweden (http://www.diamyd.com), has performed a wide range of Good Laboratory Practice (GLP)-compliant animal safety studies to support the clinical use of alum-formulated recombinant human GAD65 prior to the safety (Phase I) study. No adverse effects other than local inflammation at the injection site have been observed in mice, rats, rabbits, marmosets, or dogs when injected with recombinant human GAD65 with or without adjuvant, for periods up to 4 weeks. GAD65 immunization induced both T- and B-cell responses without inducing insulitis, diabetes, or neurological abnormalities (Bieg *et al.*, 1999; Plesner *et al.*, 1998). There are so far no reports to our knowledge of behavioral or neurological conditions arising after raising antibodies to GAD65 in mice, rats, sheep, marmosets, or goats.

Several studies were conducted with the unformulated rhGAD65 Bulk Product from Diamyd Medical AB to explore (1) acute toxicity (rat and mouse subcutaneous and intravenous injections); (2) repeat dose toxicity by injecting rats subcutaneously for 28 days or marmosets injected for 28 days and then followed for symptoms; (3) local tolerance by intradermal injections in rabbits; and (4) safety pharmacology using Irwin screen in mice and cardiovascular effects in dogs after intravenous injection of GAD65. No adverse events were observed and all studies are reported to regulatory agencies.

In addition, Ames test and mouse lymphoma assays were negative and the mouse micronucleus test showed no indication of mutagenic or potentially carcinogenic effects. Evaluation of the components showed very wide safety margins between the toxic dose levels of the ingredients and the human doses that would result from exposure to alum-formulated rhGAD65. A total of eight separate studies have supported immunological safety after administration of alum-formulated recombinant human GAD65 (from Diamyd Medical AB). These studies show that treatment with alum-formulated GAD65 had (1) no inductive or immunomodulatory impact on cellular or humoral responses to ovalbumin in the mouse; (2) no inductive or immunomodulatory impact in rodent models for arthritis or multiple sclerosis; (3) no impact on influenza infection in a mouse model; and (4) no effect on ovalbumin (OVA) sensitization for allergy. Taken together, these preclinical studies did not reveal any evidence for adverse effects after treatment with alum-formulated recombinant human GAD65.

In a clinical Phase 1 study conducted in the UK by Diamyd Medical AB, 16 healthy male Caucasian volunteers who were non-DR3-DQ2 and non-DR4-DQ8 received single subcutaneous administrations of unformulated recombinant human GAD65. This rising, single-dose, double-blind study, which also included eight volunteers who received placebo, was conducted to assess the safety and tolerability after subcutaneous administration of separate and ascending dose levels from 20 to a maximum of 500 µg of unformulated recombinant human GAD65 per volunteer. No significant treatment-related adverse clinical effects were seen at any dose-level, and no GAD65, insulin, or IA-2 autoantibodies were induced in any subject. Accordingly, it was concluded that this GAD65 treatment was clinically safe and well tolerated.

6. PHASE II GAD65 CLINICAL STUDIES

A wide dose range of alum-formulated recombinant human GAD65 was proposed for clinical investigation. Four patient groups, each representing fivefold dose increments of 4, 20, 100, and 500 µg/patient, were studied. The lowest dose (4 μg) was intended as a "No Effect" dose level. The next higher dose (20 μg) was representative of a "low" vaccine or immunomodulant dose level that would be appropriate for a highly immunogenic protein. The fivefold higher next dose level (100 μg) represents a commonly used vaccine dose level appropriate for moderately immunogenic proteins. The top dose level (500 μg) either represents a high vaccine dose level required for a poorly immunogenic protein or, alternatively, may be required to induce a strong immune response for GAD65 tolerization.

Alum was selected as the adjuvant for formulation with recombinant human GAD65 for clinical use for the following reasons: (1) repeat intraperitoneal administrations of 50 μg recombinant human GAD65 formulated in incomplete Freund's Adjuvans (IFA) induced active tolerance in NOD mice (Kaufman et al., 1993); (2) alum is used in several commercial vaccines (e.g., Diphteria, Tetanus and Pertussis (DTP)) and is the only adjuvant currently approved by the FDA; and (3) aluminum salts are recognized as preferentially inducing humoral rather than cellular immune responses. The use of alum in the formulation was to change the autoimmune response away from a cellular toward a humoral response to GAD65 in order to minimize the possibility of promoting cell-mediated beta-cell destruction.

The purpose of the Phase II study was to evaluate if alum-formulated human recombinant GAD65 was safe and did not compromise beta cell function (Agardh et al., 2005). A severe adverse effect by alum-formulated GAD65 would be if the residual beta cell function would be lost in an accelerated manner. Our study was conducted as a randomized, double-blind, placebo-controlled, dose-escalation clinical trial in a total of 47 LADA patients who received either placebo or 4, 20, 100, or 500 μg alum-GAD65 subcutaneously at weeks 1 and 4. Safety evaluations, including neurology, beta cell function tests, diabetes status assessment, hematology, biochemistry, and cellular and humoral immunological markers, were repeatedly assessed over 24 weeks.

None of the patients had significant study-related adverse events (Agardh et al., 2005). Fasting c-peptide levels at 24 weeks were increased compared with placebo in the 20-μg but not in the other dose groups. In addition, both fasting and stimulated c-peptide levels increased from baseline to 24 weeks in the 20-μg dose group. GAD65A log levels increased in response to 500 μg. The $CD4^+$–$CD25+$/$CD4^+$–$CD25-$ cell ratio increased at 24 weeks in the 20-μg group. No sudden increase in Hb A_{1c} or plasma glucose or decrease in beta cell function was observed in any of the dose groups. These positive findings for clinical safety further support the clinical development of alum-GAD65 as a therapeutic to prevent autoimmune diabetes.

The effect of recombinant human GAD65 administration on the GAD65A epitope recognition was tested using GAD65-specific recombinant Fab and GAD65/67 fusion proteins (Bekris et al., 2007). Overall, minor changes in the epitope pattern were observed using either approach. Only in the 500-μg dosage group were there increased GAD65Ab levels associated with a significant increase in the binding to a conformational epitope located at the middle part of GAD65. Our data suggest that the apparent beneficial effects of 20 μg alum-formulated recombinant human GAD65 (Agardh et al., 2005) is not explained by changes in the GAD65Ab epitope pattern.

As alum-formulated GAD65 was safe and seemed to improve residual insulin secretion in a Phase II clinical trial in LADA patients, a second trial assessed the safety and efficacy of alum-GAD65 to reverse T1D in 10–18-year-old recent onset patients (Casas et al., 2007). A total of 70 T1D patients were recruited within 18 months from diagnosis, with fasting C-peptide levels greater than 0.10 pmol/ml and positive for GAD65A. The T1D children were randomly assigned to 20 μg alum-formulated GAD65 or placebo administered subcutaneously at day 1 and 1 month later. Both fasting serum C-peptide and stimulated C-peptide secretion as area under the curve (AUC) declined less in the alum-GAD65 compared to the placebo group. Adverse events were minor, not study related, and did not differ between GAD-alum and placebo groups.

7. ONGOING AND FUTURE CLINICAL STUDIES

A Phase III clinical trial with 20 μg alum-formulated GAD65 is in progress. The study will comprise 600 T1D patients 10–18 years of age with a T1D duration of 3 months. The study will be a randomized, double-blind, placebo-controlled, multicenter study using a 20 μg GAD65 administered by subcutaneous injection on two occasions 4 weeks apart. The patients will be assessed for eligibility at the screening visit within 2 weeks prior to the first injection. On the second visit patients eligible for the study will be randomized to one of two treatment groups. One treatment group will be injected subcutaneously with placebo, and the second treatment group will be injected subcutaneously with 20 μg alum-formulated GAD65. Four weeks later the patients will receive a second ("booster") injection of the same dosage.

The patients will be followed for a total study period of 24 months comprising a main study period of 12 months and a follow-up study period of 12 months. The primary efficacy endpoint is the change in C-peptide concentration after a Mixed Meal Tolerance Test (0–120 min), AUC from baseline to month 12 after the first injection. Adverse events will be determined, which is one of the primary objectives of the study. Other studies with alum-formulated GAD65 are also planned; for example, it will be explored if additional injections of alum-GAD65 will be beneficial.

8. CONCLUDING REMARKS

GAD65 is a major autoantigen in T1D. Autoreactivity to GAD65 is particularly prominent in teenagers, young adults, and among adults and differs from insulin and IA-2 autoreactivities, which are more common in infants and children. Our understanding of the role of GAD65 in T1D etiology and pathogenesis is improving continuously. Studies such as the Natural History Study of TrialNet (http://www2.diabetestrialnet.org/) and the TEDDY study (TEDDY, 2007) show promise to uncover the spontaneous appearance of GAD65A and eventually, once standardized, cellular responses as well. SIT with alum-formulated human recombinant GAD65 has been attempted in GAD65A-positive patients. We conclude from the first safety Phase II clinical trial in LADA patients that there were no adverse events and a beneficial effect on residual beta-cell function (Agardh et al., 2005). Similar results were observed in a second safety Phase II clinical trial in T1D children and teenagers (Casas et al., 2007). Taken together, current experience in randomized, double-blind, placebo-controlled clinical trials suggest that it is possible to use recombinant human GAD65 in subcutaneous immunotherapy (SCIT) to modulate islet autoreactivity in humans. Further controlled clinical trials are needed to uncover the mechanisms by which subcutaneous injections of recombinant human GAD65 may alter GAD65 autoreactivity.

ACKNOWLEDGMENTS

We thank Gustavo Fenalti and Merrill Rowley for the assistance in preparing Figures 3.3 and 3.4. R. U. was supported by the EU Marie Curie ToK grant and a grant from the Kungliga Fysiografiska Sällskapet. Å. L. was supported by the Robert H. Williams Endowed Chair at the University of Washington. Studies in the authors' laboratories were supported by the National Institutes of Health (Grant DK26190, DK53004), Swedish Diabetes Association, Knut and Alice Wallenberg Foundation, European Foundation for the Study of Diabetes (EFSD), Juvenile Diabetes Research Foundation, and the UMAS Fund.

REFERENCES

Abraham, R. S., Kudva, Y. C., Wilson, S. B., Strominger, J. L., and David, C. S. (2000). Co-expression of HLA DR3 and DQ8 results in the development of spontaneous insulitis and loss of tolerance to GAD65 in transgenic mice. *Diabetes* **49**, 548.

Agardh, C. D., Cilio, C. M., Lethagen, A., Lynch, K., Leslie, R. D., Palmer, M., Harris, R. A., Robertson, J. A., and Lernmark, A. (2005). Clinical evidence for the safety of GAD65 immunomodulation in adult-onset autoimmune diabetes. *J. Diab. Complicat.* **19**, 238.

Ankelo, M., Westerlund, A., Blomberg, K., Knip, M., Ilonen, J., and Hinkkanen, A. E. (2007). Time-resolved immunofluorometric dual-label assay for simultaneous detection of auto-antibodies to GAD65 and IA-2 in children with type 1 diabetes. *Clin. Chem.* **53**, 472.

Arif, S., Tree, T. I., Astill, T. P., Tremble, J. M., Bishop, A. J., Dayan, C. M., Roep, B. O., and Peakman, M. (2004). Autoreactive T cell responses show proinflammatory polarization in diabetes but a regulatory phenotype in health. *J. Clin. Invest.* **113**, 451.

Atkinson, M. A., Kaufman, D. L., Campbell, L., Gibbs, K. A., Shah, S. C., Bu, D. F., Erlander, M. G., Tobin, A. J., and Maclaren, N. K. (1992). Response of peripheral-blood mononuclear cells to glutamate decarboxylase in insulin-dependent diabetes. *Lancet* **339**, 458.

Bach, J. F. (2003). Regulatory T cells under scrutiny. *Nat. Rev. Immunol.* **3**, 189.

Baekkeskov, S., Kanatsuna, T., Klareskog, L., Nielsen, D. A., Peterson, P. A., Rubenstein, A. H., Steiner, D. F., and Lernmark, A. (1981). Expression of major histocompatibility antigens on pancreatic islet cells. *Proc. Natl. Acad. Sci. USA* **78**, 6456.

Baekkeskov, S., Nielsen, J. H., Marner, B., Bilde, T., Ludvigsson, J., and Lernmark, A. (1982). Autoantibodies in newly diagnosed diabetic children immunoprecipitate human pancreatic islet cell proteins. *Nature* **298**, 167.

Baekkeskov, S., Aanstoot, H. J., Christgau, S., Reetz, A., Solimena, M., Cascalho, M., Folli, F., Richter-Olesen, H., and De Camilli, P. (1990). Identification of the 64K autoantigen in insulin-dependent diabetes as the GABA-synthesizing enzyme glutamic acid decarboxylase. *Nature* **347**, 151.

Banga, J. P., Moore, J. K., Duhindan, N., Madec, A. M., van Endert, P. M., Orgiazzi, J., and Endl, J. (2004). Modulation of antigen presentation by autoreactive B cell clones specific for GAD65 from a type I diabetic patient. *Clin. Exp. Immunol.* **135**, 74.

Bekris, L. M., Jensen, R. A., Lagerquist, E., Hall, T. R., Agardh, C. D., Cilio, C. M., Lethagen, A. L., Lernmark, A., Robertson, J. A., and Hampe, C. S. (2007). GAD65 autoantibody epitopes in adult patients with latent autoimmune diabetes following GAD65 vaccination. *Diabet. Med.* **24**, 521.

Bieg, S., Moller, C., Olsson, T., and Lernmark, A. (1997). The lymphopenia (lyp) gene controls the intrathymic cytokine ratio in congenic BioBreeding rats. *Diabetologia* **40**, 786.

Bieg, S., Hanlon, C., Hampe, C. S., Benjamin, D., and Mahoney, C. P. (1999). GAD65 and insulin B chain peptide (9–23) are not primary autoantigens in the type 1 diabetes syndrome of the BB rat. *Autoimmunity* **31**, 15.

Bingley, P. J., Bonifacio, E., and Mueller, P. W. (2003). Diabetes Antibody Standardization Program: First assay proficiency evaluation. *Diabetes* **52**, 1128.

Bjork, E., Velloso, L. A., Kampe, O., and Karlsson, F. A. (1994). GAD autoantibodies in IDDM, stiff-man syndrome, and autoimmune polyendocrine syndrome type I recognize different epitopes. *Diabetes* **43**, 161.

Blancou, P., Mallone, R., Martinuzzi, E., Severe, S., Pogu, S., Novelli, G., Bruno, G., Charbonnel, B., Dolz, M., Chaillous, L., van Endert, P., and Bach, J. M. (2007). Immunization of HLA class I transgenic mice identifies autoantigenic epitopes eliciting dominant responses in type 1 diabetes patients. *J. Immunol.* **178**, 7458.

Bonifacio, E., Lampasona, V., Bernasconi, L., and Ziegler, A. G. (2000). Maturation of the humoral autoimmune response to epitopes of GAD in preclinical childhood type 1 diabetes. *Diabetes* **49**, 202.

Bottazzo, G. F., and Lendrum, R. (1976). Separate autoantibodies to human pancreatic glucagon and somatostatin cells. *Lancet* **2**, 873.

Bottazzo, G. F., Florin-Christensen, A., and Doniach, D. (1974). Islet-cell antibodies in diabetes mellitus with autoimmune polyendocrine deficiencies. *Lancet* **2**, 1279.

Bresson, D., and von Herrath, M. (2007). Moving towards efficient therapies in type 1 diabetes: To combine or not to combine? *Autoimmun. Rev.* **6**, 315.

Brilliant, M. H., Szabo, G., Katarova, Z., Kozak, C. A., Glaser, T. M., Greenspan, R. J., and Housman, D. E. (1990). Sequences homologous to glutamic acid decarboxylase cDNA are present on mouse chromosomes 2 and 10. *Genomics* **6**, 115.

Brooking, H., Ananieva-Jordanova, R., Arnold, C., Amoroso, M., Powell, M., Betterle, C., Zanchetta, R., Furmaniak, J., and Smith, B. R. (2003). A sensitive non-isotopic assay for GAD65 autoantibodies. *Clin. Chim. Acta* **331**, 55.

Brusko, T., Wasserfall, C., McGrail, K., Schatz, R., Viener, H. L., Schatz, D., Haller, M., Rockell, J., Gottlieb, P., Clare-Salzler, M., and Atkinson, M. (2007). No alterations in the frequency of FOXP3+ regulatory T-cells in type 1 diabetes. *Diabetes* **56**, 604.

Bu, D. F., Erlander, M. G., Hitz, B. C., Tillakaratne, N. J., Kaufman, D. L., Wagner-McPherson, C. B., Evans, G. A., and Tobin, A. J. (1992). Two human glutamate decarboxylases, 65-kDa GAD and 67-kDa GAD, are each encoded by a single gene. *Proc. Natl. Acad. Sci. USA* **89**, 2115.

Burbelo, P. D., Groot, S., Dalakas, M. C., and Iadarola, M. J. (2008). High definition profiling of autoantibodies to glutamic acid decarboxylases GAD65/GAD67 in stiff-person syndrome. *Biochem. Biophys. Res. Commun.* **366**, 1.

Casas, R., Hedman, M., Faresjo, M., and Ludvigsson, J. (2007). *Diabetes* 67th Annual Scientific Sessions, 1242.

Chessler, S. D., and Lernmark, A. (2000). Alternative splicing of GAD67 results in the synthesis of a third form of glutamic-acid decarboxylase in human islets and other non-neural tissues. *J. Biol. Chem.* **275**, 5188.

Couper, J. J., Harrison, L. C., Aldis, J. J., Colman, P. G., Honeyman, M. C., and Ferrante, A. (1998). IgG subclass antibodies to glutamic acid decarboxylase and risk for progression to clinical insulin-dependent diabetes. *Hum. Immunol.* **59**, 493.

Danke, N. A., Yang, J., Greenbaum, C., and Kwok, W. W. (2005). Comparative study of GAD65-specific CD4$^+$ T cells in healthy and type 1 diabetic subjects. *J. Autoimmun.* **25**, 303.

DiLorenzo, T. P., and Serreze, D. V. (2005). The good turned ugly: Immunopathogenic basis for diabetogenic CD8$^+$ T cells in NOD mice. *Immunol. Rev.* **204**, 250.

DiLorenzo, T. P., Peakman, M., and Roep, B. O. (2007). Translational mini-review series on type 1 diabetes: Systematic analysis of T cell epitopes in autoimmune diabetes. *Clin. Exp. Immunol.* **148**, 1.

Dirkx, R., Jr, Thomas, A., Li, L., Lernmark, A., Sherwin, R. S., De Camilli, P., and Solimena, M. (1995). Targeting of the 67-kDa isoform of glutamic acid decarboxylase to intracellular organelles is mediated by its interaction with the NH2-terminal region of the 65-kDa isoform of glutamic acid decarboxylase. *J. Biol. Chem.* **270**, 2241.

Dobersen, M. J., Scharff, J. E., Ginsberg-Fellner, F., and Notkins, A. L. (1980). Cytotoxic autoantibodies to beta cells in the serum of patients with insulin-dependent diabetes mellitus. *N. Engl. J. Med.* **303**, 1493.

Eisenbarth, G. S., Moriyama, H., Robles, D. T., Liu, E., Yu, L., Babu, S., Redondo, M. J., Gottlieb, P., Wegmann, D., and Rewers, M. (2002). Insulin autoimmunity: prediction/precipitation/prevention type 1A diabetes. *Autoimmun. Rev.* **1**, 139.

Ellis, R. J., Varela-Calvino, R., Tree, T. I., and Peakman, M. (2005). HLA Class II molecules on haplotypes associated with type 1 diabetes exhibit similar patterns of binding affinities for coxsackievirus P2C peptides. *Immunology* **116**, 337.

Endl, J., Otto, H., Jung, G., Dreisbusch, B., Donie, F., Stahl, P., Elbracht, R., Schmitz, G., Meinl, E., Hummel, M., Ziegler, A. G., Wank, R., et al. (1997). Identification of naturally processed T cell epitopes from glutamic acid decarboxylase presented in the context of HLA-DR alleles by T lymphocytes of recent onset IDDM patients. *J. Clin. Invest.* **99**, 2405.

Falorni, A., Grubin, C. E., Takei, I., Shimada, A., Kasuga, A., Maruyama, T., Ozawa, Y., Kasatani, T., Saruta, T., Li, L., and Lernmark, A. (1994). Radioimmunoassay detects the frequent occurrence of autoantibodies to the Mr 65,000 isoform of glutamic acid decarboxylase in Japanese insulin-dependent diabetes. *Autoimmunity* **19**, 113.

Fenalti, G., Hampe, C. S., O'Connor, K., Banga, J. P., Mackay, I. R., Rowley, M. J., and El-Kabbani, O. (2007a). COOH-terminal clustering of autoantibody and T-cell determinants on the structure of GAD65 provide insights into the molecular basis of autoreactivity. *Mol. Immunol.* **44**, 1178.

Fenalti, G., Law, R. H., Buckle, A. M., Langendorf, C., Tuck, K., Rosado, C. J., Faux, N. G., Mahmood, K., Hampe, C. S., Banga, J. P., Wilce, M., Schmidberger, J., et al. (2007b). GABA production by glutamic acid decarboxylase is regulated by a dynamic catalytic loop. Nat. Struct. Mol. Biol. **14,** 280.

Fenalti, G., Hampe, C. S., Arafat, Y., Law, R. H., Banga, J. P., Mackay, I. R., Whisstock, J. C., Buckle, A. M., and Rowley, M. J. (2008). COOH-terminal clustering of autoantibody and T-cell determinants on the structure of GAD65 provide insights into the molecular basis of autoreactivity. Diabetes **57,** 1293.

Gianani, R., Pugliese, A., Bonner-Weir, S., Shiffrin, A. J., Soeldner, J. S., Erlich, H., Awdeh, Z., Alper, C. A., Jackson, R. A., and Eisenbarth, G. S. (1992). Prognostically significant heterogeneity of cytoplasmic islet cell antibodies in relatives of patients with type I diabetes. Diabetes **41,** 347.

Gotter, J., and Kyewski, B. (2004). Regulating self-tolerance by deregulating gene expression. Curr. Opin. Immunol. **16,** 741.

Graham, J., Hagopian, W. A., Kockum, I., Li, L. S., Sanjeevi, C. B., Lowe, R. M., Schaefer, J. B., Zarghami, M., Day, H. L., Landin-Olsson, M., Palmer, J. P., Janer-Villanueva, M., et al. (2002). Genetic effects on age-dependent onset and islet cell autoantibody markers in type 1 diabetes. Diabetes **51,** 1346.

Greening, J. E., Tree, T. I., Kotowicz, K. T., van Halteren, A. G., Roep, B. O., Klein, N. J., and Peakman, M. (2003). Processing and presentation of the islet autoantigen GAD by vascular endothelial cells promotes transmigration of autoreactive T-cells. Diabetes **52,** 717.

Grubin, C. E., Daniels, T., Toivola, B., Landin-Olsson, M., Hagopian, W. A., Li, L., Karlsen, A. E., Boel, E., Michelsen, B., and Lernmark, A. (1994). A novel radioligand binding assay to determine diagnostic accuracy of isoform-specific glutamic acid decarboxylase antibodies in childhood IDDM. Diabetologia **37,** 344.

Gutcher, I., and Becher, B. (2007). APC-derived cytokines and T cell polarization in autoimmune inflammation. J. Clin. Invest. **117,** 1119.

Hagopian, W. A., Lernmark, A., Rewers, M. J., Simell, O. G., She, J. X., Ziegler, A. G., Krischer, J. P., and Akolkar, B. (2006). TEDDY–The Environmental Determinants of Diabetes in the Young: an observational clinical trial. Ann. N. Y. Acad. Sci. **1079,** 320.

Hampe, C. S., Ortqvist, E., Rolandsson, O., Landin-Olsson, M., Torn, C., Agren, A., Persson, B., Schranz, D. B., and Lernmark, A. (1999). Species-specific autoantibodies in type 1 diabetes. J. Clin. Endocrinol. Metab. **84,** 643.

Hampe, C. S., Hammerle, L. P., Bekris, L., Ortqvist, E., Kockum, I., Rolandsson, O., Landin-Olsson, M., Torn, C., Persson, B., and Lernmark, A. (2000). Recognition of glutamic acid decarboxylase (GAD) by autoantibodies from different GAD antibody-positive phenotypes. J. Clin. Endocrinol. Metab. **85,** 4671.

Hawa, M. I., Fava, D., Medici, F., Deng, Y. J., Notkins, A. L., De Mattia, G., and Leslie, R. D. (2000). Antibodies to IA-2 and GAD65 in type 1 and type 2 diabetes: Isotype restriction and polyclonality. Diabetes Care **23,** 228.

Holm, B. C., Svensson, J., Akesson, C., Arvastsson, J., Ljungberg, J., Lynch, K., Ivarsson, S. A., Lernmark, A., and Cilio, C. M. (2006). Evidence for immunological priming and increased frequency of CD4$^+$ CD25+ cord blood T cells in children born to mothers with type 1 diabetes. Clin. Exp. Immunol. **146,** 493.

Hoppu, S., Harkonen, T., Ronkainen, M. S., Akerblom, H. K., and Knip, M. (2004). IA-2 antibody epitopes and isotypes during the prediabetic process in siblings of children with type 1 diabetes. J. Autoimmun. **23,** 361.

Huen, A. H., Haneda, M., Freedman, Z., Lernmark, A., and Rubenstein, A. H. (1983). Quantitative determination of islet cell surface antibodies using 125I-protein A. Diabetes **32,** 460.

Hummel, M., Bonifacio, E., Naserke, H. E., and Ziegler, A. G. (2002). Elimination of dietary gluten does not reduce titers of type 1 diabetes-associated autoantibodies in high-risk subjects. *Diabetes Care* **25,** 1111.
Hyoty, H., and Taylor, K. W. (2002). The role of viruses in human diabetes. *Diabetologia* **45,** 1353.
In't Veld, P., Lievens, D., De Grijse, J., Ling, Z., Van der Auwera, B., Pipeleers-Marichal, M., Gorus, F., and Pipeleers, D. (2007). Screening for insulitis in adult autoantibody-positive organ donors. *Diabetes* **56,** 2400.
Itoh, N., Hanafusa, T., Miyazaki, A., Miyagawa, J., Yamagata, K., Yamamoto, K., Waguri, M., Imagawa, A., Tamura, S., Inada, M., Sugawara, K., Kobayashi, T., et al. (1993). Mononuclear cell infiltration and its relation to the expression of major histocompatibility complex antigens and adhesion molecules in pancreas biopsy specimens from newly diagnosed insulin-dependent diabetes mellitus patients. *J. Clin. Invest.* **92,** 2313.
James, E. A., Moustakas, A. K., Berger, D., Huston, L., Papadopoulos, G. K., and Kwok, W. W. (2008). Definition of the peptide binding motif within DRB1*1401 restricted epitopes by peptide competition and structural modeling. *Mol. Immunol.* **45,** 2651.
Jansen, A., van Hagen, M., and Drexhage, H. A. (1995). Defective maturation and function of antigen-presenting cells in type 1 diabetes. *Lancet* **345,** 491.
Jaume, J. C., Parry, S. L., Madec, A. M., Sonderstrup, G., and Baekkeskov, S. (2002). Suppressive effect of glutamic acid decarboxylase 65-specific autoimmune B lymphocytes on processing of T cell determinants located within the antibody epitope. *J. Immunol.* **169,** 665.
Jones, D. B., and Crosby, I. (1996). Proliferative lymphocyte responses to virus antigens homologous to GAD65 in IDDM. *Diabetologia* **39,** 1318.
Karlsen, A. E., Hagopian, W. A., Grubin, C. E., Dube, S., Disteche, C. M., Adler, D. A., Barmeier, H., Mathewes, S., Grant, F. J., Foster, D., et al. (1991). Cloning and primary structure of a human islet isoform of glutamic acid decarboxylase from chromosome 10. *Proc. Natl. Acad. Sci. USA* **88,** 8337.
Kaufman, D. L., Clare-Salzler, M., Tian, J., Forsthuber, T., Ting, G. S., Robinson, P., Atkinson, M. A., Sercarz, E. E., Tobin, A. J., and Lehmann, P. V. (1993). Spontaneous loss of T-cell tolerance to glutamic acid decarboxylase in murine insulin-dependent diabetes. *Nature* **366,** 69.
Kaufman, D. L., Tisch, R., Sarvetnick, N., Chatenoud, L., Harrison, L. C., Haskins, K., Quinn, A., Sercarz, E., Singh, B., von Herrath, M., Wegmann, D., Wen, L., et al. (2001). Report from the 1st International NOD Mouse T-Cell Workshop and the follow-up miniworkshop. *Diabetes* **50,** 2459.
Kobayashi, T., Tanaka, S., Okubo, M., Nakanishi, K., Murase, T., and Lernmark, A. (2003). Unique epitopes of glutamic acid decarboxylase autoantibodies in slowly progressive type 1 diabetes. *J. Clin. Endocrinol. Metab.* **88,** 4768.
Korn, T., and Oukka, M. (2007). Dynamics of antigen-specific regulatory T-cells in the context of autoimmunity. *Semin. Immunol.* **19,** 272.
Kotani, R., Nagata, M., Moriyama, H., Nakayama, M., Yamada, K., Chowdhury, S. A., Chakrabarty, S., Jin, Z., Yasuda, H., and Yokono, K. (2002). Detection of GAD65-reactive T-Cells in type 1 diabetes by immunoglobulin-free ELISPOT assays. *Diabetes Care* **25,** 1390.
Law, R. H., Rowley, M. J., Mackay, I. R., and Corner, B. (1998). Expression in *Saccharomyces cerevisiae* of antigenically and enzymatically active recombinant glutamic acid decarboxylase. *J. Biotechnol.* **61,** 57.
Lee, S. H., Lee, H., Park, J. S., Choi, H., Han, K. Y., Seo, H. S., Ahn, K. Y., Han, S. S., Cho, Y., Lee, K. H., and Lee, J. (2007). A novel approach to ultrasensitive diagnosis using supramolecular protein nanoparticles. *FASEB J.* **21,** 1324.

Lendrum, R., Walker, G., Cudworth, A. G., Theophanides, C., Pyke, D. A., Bloom, A., and Gamble, D. R. (1976). Islet-cell antibodies in diabetes mellitus. *Lancet* **2**, 1273.

Lernmark, A., and Agardh, C. D. (2005). Immunomodulation with human recombinant autoantigens. *Trends Immunol.* **26**, 608.

Lernmark, A., and Baekkeskov, S. (1981). Islet cell antibodies-theoretical and practical implications. *Diabetologia* **21**, 431.

Lernmark, A., Freedman, Z. R., Hofmann, C., Rubenstein, A. H., Steiner, D. F., Jackson, R. L., Winter, R. J., and Traisman, H. S. (1978). Islet-cell-surface antibodies in juvenile diabetes mellitus. *N. Engl. J. Med.* **299**, 375.

MacCuish, A. C., Irvine, W. J., Barnes, E. W., and Duncan, L. J. (1974). Antibodies to pancreatic islet cells in insulin-dependent diabetics with coexistent autoimmune disease. *Lancet* **2**, 1529.

Maclaren, N. K., Huang, S. W., and Fogh, J. (1975). Antibody to cultured human insulinoma cells in insulin-dependent diabetes. *Lancet* **1**, 997.

Madec, A. M., Rousset, F., Ho, S., Robert, F., Thivolet, C., Orgiazzi, J., and Lebecque, S. (1996). Four IgG anti-islet human monoclonal antibodies isolated from a type 1 diabetes patient recognize distinct epitopes of glutamic acid decarboxylase 65 and are somatically mutated. *J. Immunol.* **156**, 3541.

Mallone, R., and Nepom, G. T. (2004). MHC Class II tetramers and the pursuit of antigen-specific T cells: Define, deviate, delete. *Clin. Immunol.* **110**, 232.

Mallone, R., and Nepom, G. T. (2005). Targeting T lymphocytes for immune monitoring and intervention in autoimmune diabetes. *Am. J. Ther.* **12**, 534.

Marshall, M. O., Hoyer, P. E., Petersen, J. S., Hejnaes, K. R., Genovese, S., Dyrberg, T., and Bottazzo, G. F. (1994). Contribution of glutamate decarboxylase antibodies to the reactivity of islet cell cytoplasmic antibodies. *J. Autoimmun.* **7**, 497.

Martinuzzi, E., Novelli, G., Scotto, M., Blancou, P., Bach, J. M., Chaillous, L., Bruno, G., Chatenoud, L., van Endert, P., and Mallone, R. (2008). The frequency and immunodominance of islet-specific CD8[+] T-cell responses change after type 1 diabetes diagnosis and treatment. *Diabetes* **57**, 1312.

Matsukawa, S., and Ueno, H. (2007). Expression of glutamate decarboxylase isoform, GAD65, in human mononuclear leucocytes: A possible implication of C-terminal end deletion by Western blot and RT-PCR study. *J. Biochem.* **142**, 633.

Mire-Sluis, A. R., Gaines Das, R., and Lernmark, A. (2000). The World Health Organization International Collaborative Study for islet cell antibodies. *Diabetologia* **43**, 1282.

Mollah, Z. U., Pai, S., Moore, C., O'Sullivan, B. J., Harrison, M. J., Peng, J., Phillips, K., Prins, J. B., Cardinal, J., and Thomas, R. (2008). Abnormal NF-kappaB function characterizes human type 1 diabetes dendritic cells and monocytes. *J. Immunol.* **180**, 3166.

Monti, P., Scirpoli, M., Rigamonti, A., Mayr, A., Jaeger, A., Bonfanti, R., Chiumello, G., Ziegler, A. G., and Bonifacio, E. (2007). Evidence for *in vivo* primed and expanded autoreactive T cells as a specific feature of patients with type 1 diabetes. *J. Immunol.* **179**, 5785.

Moody, A. J., Hejnaes, K. R., Marshall, M. O., Larsen, F. S., Boel, E., Svendsen, I., Mortensen, E., and Dyrberg, T. (1995). Isolation by anion-exchange of immunologically and enzymatically active human islet glutamic acid decarboxylase 65 overexpressed in Sf9 insect cells. *Diabetologia* **38**, 14.

Nagata, M., Kotani, R., Moriyama, H., Yokono, K., Roep, B. O., and Peakman, M. (2004). Detection of autoreactive T cells in type 1 diabetes using coded autoantigens and an immunoglobulin-free cytokine ELISPOT assay: Report from the fourth immunology of diabetes society T cell workshop. *Ann. N. Y. Acad. Sci.* **1037**, 10.

Namchuk, M., Lindsay, L., Turck, C. W., Kanaani, J., and Baekkeskov, S. (1997). Phosphorylation of serine residues 3, 6, 10, and 13 distinguishes membrane anchored from soluble glutamic acid decarboxylase 65 and is restricted to glutamic acid decarboxylase 65alpha. *J. Biol. Chem.* **272**, 1548.

Nepom, G. T., Lippolis, J. D., White, F. M., Masewicz, S., Marto, J. A., Herman, A., Luckey, C. J., Falk, B., Shabanowitz, J., Hunt, D. F., Engelhard, V. H., and Nepom, B. S. (2001). Identification and modulation of a naturally processed T cell epitope from the diabetes-associated autoantigen human glutamic acid decarboxylase 65 (hGAD65). *Proc. Natl. Acad. Sci. USA* **98**, 1763.

Nurieva, R., Yang, X. O., Martinez, G., Zhang, Y., Panopoulos, A. D., Ma, L., Schluns, K., Tian, Q., Watowich, S. S., Jetten, A. M., and Dong, C. (2007). Essential autocrine regulation by IL-21 in the generation of inflammatory T cells. *Nature* **448**, 480.

Oak, S., Gilliam, L. K., Landin-Olsson, M., Torn, C., Kockum, I., Pennington, C. R., Rowley, M. J., Christie, M. R., Banga, J. P., and Hampe, C. S. (2008). The lack of anti-idiotypic antibodies, not the presence of the corresponding autoantibodies to glutamate decarboxylase, defines type 1 diabetes. *Proc. Natl. Acad. Sci. USA* **105**, 5471.

Oling, V., Marttila, J., Ilonen, J., Kwok, W. W., Nepom, G., Knip, M., Simell, O., and Reijonen, H. (2005). GAD65- and proinsulin-specific CD4[+] T-cells detected by MHC class II tetramers in peripheral blood of type 1 diabetes patients and at-risk subjects. *J. Autoimmun.* **25**, 235.

Ott, P. A., Dittrich, M. T., Herzog, B. A., Guerkov, R., Gottlieb, P. A., Putnam, A. L., Durinovic-Bello, I., Boehm, B. O., Tary-Lehmann, M., and Lehmann, P. V. (2004). T cells recognize multiple GAD65 and proinsulin epitopes in human type 1 diabetes, suggesting determinant spreading. *J. Clin. Immunol.* **24**, 327.

Ou, D., Mitchell, L. A., Metzger, D. L., Gillam, S., and Tingle, A. J. (2000). Cross-reactive rubella virus and glutamic acid decarboxylase (65 and 67) protein determinants recognised by T cells of patients with type I diabetes mellitus. *Diabetologia* **43**, 750.

Padoa, C. J., Banga, J. P., Madec, A. M., Ziegler, M., Schlosser, M., Ortqvist, E., Kockum, I., Palmer, J., Rolandsson, O., Binder, K. A., Foote, J., Luo, D., *et al.* (2003). Recombinant Fabs of human monoclonal antibodies specific to the middle epitope of GAD65 inhibit type 1 diabetes-specific GAD65Abs. *Diabetes* **52**, 2689.

Palmer, J. P., Hampe, C. S., Chiu, H., Goel, A., and Brooks-Worrell, B. M. (2005). Is latent autoimmune diabetes in adults distinct from type 1 diabetes or just type 1 diabetes at an older age? *Diabetes* **54**, S62.

Panina-Bordignon, P., Lang, R., van Endert, P. M., Benazzi, E., Felix, A. M., Pastore, R. M., Spinas, G. A., and Sinigaglia, F. (1995). Cytotoxic T cells specific for glutamic acid decarboxylase in autoimmune diabetes. *J. Exp. Med.* **181**, 1923.

Papouchado, M. L., Ermacora, M. R., and Poskus, E. (1996). Detection of glutamic acid decarboxylase (GAD) autoantibodies by indirect immunofluorescence using CHO cells expressing recombinant human GAD65. *J. Autoimmun.* **9**, 689.

Petersen, J. S., Hejnaes, K. R., Moody, A., Karlsen, A. E., Marshall, M. O., Hoier-Madsen, M., Boel, E., Michelsen, B. K., and Dyrberg, T. (1994a). Detection of GAD65 antibodies in diabetes and other autoimmune diseases using a simple radioligand assay. *Diabetes* **43**, 459.

Petersen, J. S., Karlsen, A. E., Markholst, H., Worsaae, A., Dyrberg, T., and Michelsen, B. (1994b). Neonatal tolerization with glutamic acid decarboxylase but not with bovine serum albumin delays the onset of diabetes in NOD mice. *Diabetes* **43**, 1478.

Petersen, J. S., Kulmala, P., Clausen, J. T., Knip, M., and Dyrberg, T. (1999). Neonatal tolerization with glutamic acid decarboxylase but not with bovine serum albumin delays the onset of diabetes in NOD mice. *Clin. Immunol.* **90**, 276.

Pihoker, C., Gilliam, L. K., Hampe, C. S., and Lernmark, A. (2005). Autoantibodies in diabetes. *Diabetes* **54**, S52.

Pitkanen, J., and Peterson, P. (2003). Autoimmune regulator: From loss of function to autoimmunity. *Genes Immun.* **4**, 12.

Plesner, A., Worsaae, A., Dyrberg, T., Gotfredsen, C., Michelsen, B. K., and Petersen, J. S. (1998). Immunization of diabetes-prone or non-diabetes-prone mice with GAD65 does not induce diabetes or islet cell pathology. *J. Autoimmun.* **11,** 335.

Plesner, A., Hampe, C. S., Daniels, T. L., Hammerle, L. P., and Lernmark, A. (2001). Preservation of enzyme activity and antigenicity after mutagenesis of the membrane anchoring domain of GAD65. *Autoimmunity* **34,** 221.

Powell, M., Prentice, L., Asawa, T., Kato, R., Sawicka, J., Tanaka, H., Petersen, V., Munkley, A., Morgan, S., Rees Smith, B., and Furmaniak, J. (1996). Glutamic acid decarboxylase autoantibody assay using 125I-labelled recombinant GAD65 produced in yeast. *Clin. Chim. Acta* **256,** 175.

Pugliese, A., Brown, D., Garza, D., Murchison, D., Zeller, M., Redondo, M. J., Diez, J., Eisenbarth, G. S., Patel, D. D., and Ricordi, C. (2001). Self-antigen-presenting cells expressing diabetes-associated autoantigens exist in both thymus and peripheral lymphoid organs. *J. Clin. Invest.* **107,** 555.

Rabinovitch, A. (1994). Immunoregulatory and cytokine imbalances in the pathogenesis of IDDM. Therapeutic intervention by immunostimulation? *Diabetes* **43,** 613.

Radillo, O., Nocera, A., Leprini, A., Barocci, S., Mollnes, T. E., Pocecco, M., Pausa, M., Valente, U., Betterle, C., and Tedesco, F. (1996). Complement-fixing islet cell antibodies in type-1 diabetes can trigger the assembly of the terminal complement complex on human islet cells and are potentially cytotoxic. *Clin. Immunol. Immunopathol.* **79,** 217.

Rajagopal, D., Bal, V., Mayor, S., George, A., and Rath, S. (2006). A role for the Hsp90 molecular chaperone family in antigen presentation to T lymphocytes via major histocompatibility complex class II molecules. *Eur. J. Immunol.* **36,** 828.

Raju, R., Foote, J., Banga, J. P., Hall, T. R., Padoa, C. J., Dalakas, M. C., Ortqvist, E., and Hampe, C. S. (2005). Analysis of GAD65 autoantibodies in Stiff-Person syndrome patients. *J. Immunol.* **175,** 7755.

Raymond, C. K., Bukowski, T., Holderman, S. D., Ching, A. F., Vanaja, E., and Stamm, M. R. (1998). Development of the methylotrophic yeast Pichia methanolica for the expression of the 65 kilodalton isoform of human glutamate decarboxylase. *Yeast* **14,** 11.

Reijonen, H., Elliott, J. F., van Endert, P., and Nepom, G. (1999). Differential presentation of glutamic acid decarboxylase 65 (GAD65) T cell epitopes among HLA-DRB1*0401-positive individuals. *J. Immunol.* **163,** 1674.

Reijonen, H., Daniels, T. L., Lernmark, A., and Nepom, G. T. (2000). GAD65-specific autoantibodies enhance the presentation of an immunodominant T-cell epitope from GAD65. *Diabetes* **49,** 1621.

Reijonen, H., Novak, E. J., Kochik, S., Heninger, A., Liu, A. W., Kwok, W. W., and Nepom, G. T. (2002). Detection of GAD65-specific T-cells by major histocompatibility complex class II tetramers in type 1 diabetic patients and at-risk subjects. *Diabetes* **51,** 1375.

Reijonen, H., Mallone, R., Heninger, A. K., Laughlin, E. M., Kochik, S. A., Falk, B., Kwok, W. W., Greenbaum, C., and Nepom, G. T. (2004). GAD65-specific CD4[+] T-cells with high antigen avidity are prevalent in peripheral blood of patients with type 1 diabetes. *Diabetes* **53,** 1987.

Richter, W., Endl, J., Eiermann, T. H., Brandt, M., Kientsch-Engel, R., Thivolet, C., Jungfer, H., and Scherbaum, W. A. (1992). GAD65-specific CD4[+] T-cells with high antigen avidity are prevalent in peripheral blood of patients with type 1 diabetes. *Proc. Natl. Acad. Sci. USA* **89,** 8467.

Richter, W., Jury, K. M., Loeffler, D., Manfras, B. J., Eiermann, T. H., Boehm, B. O. (1995). Immunoglobulin variable gene analysis of human autoantibodies reveals antigen-driven immune response to glutamate decarboxylase in type 1 diabetes mellitus. *Eur. J. Immunol.* **25,** 1703.

Roep, B. O., Atkinson, M. A., van Endert, P. M., Gottlieb, P. A., Wilson, S. B., and Sachs, J. A. (1999). Autoreactive T cell responses in insulin-dependent (Type 1) diabetes mellitus. Report of the first international workshop for standardization of T cell assays. *J. Autoimmun.* **13**, 267.

Rolandsson, O., Hagg, E., Nilsson, M., Hallmans, G., Mincheva-Nilsson, L., and Lernmark, A. (2001). Prediction of diabetes with body mass index, oral glucose tolerance test and islet cell autoantibodies in a regional population. *J. Intern. Med.* **249**, 279.

Ronkainen, M. S., Hoppu, S., Korhonen, S., Simell, S., Veijola, R., Ilonen, J., Simell, O., and Knip, M. (2006). Early epitope- and isotype-specific humoral immune responses to GAD65 in young children with genetic susceptibility to type 1 diabetes. *Eur. J. Endocrinol.* **155**, 633.

Sai, P., Rivereau, A. S., Granier, C., Haertle, T., and Martignat, L. (1996). Immunization of non-obese diabetic (NOD) mice with glutamic acid decarboxylase4-derived peptide 524–543 reduces cyclophosphamide-accelerated diabetes. *Clin. Exp. Immunol.* **105**, 330.

Sanda, S., Roep, B. O., and von Herrath, M. (2008). Islet antigen specific IL-10+ immune responses but not CD4$^+$CD25+FoxP3+ cells at diagnosis predict glycemic control in type 1 diabetes. *Clin. Immunol.* **127**, 138.

Sanjeevi, C. B., Lybrand, T. P., DeWeese, C., Landin-Olsson, M., Kockum, I., Dahlquist, G., Sundkvist, G., Stenger, D., and Lernmark, A. (1995). Polymorphic amino acid variations in HLA-DQ are associated with systematic physical property changes and occurrence of IDDM. Members of the Swedish Childhood Diabetes Study. *Diabetes* **44**, 125.

Sarugeri, E., Dozio, N., Meschi, F., Pastore, M. R., and Bonifacio, E. (2001). T cell responses to type 1 diabetes related peptides sharing homologous regions. *J. Mol. Med.* **79**, 213.

Schlosser, M., Banga, J. P., Madec, A. M., Binder, K. A., Strebelow, M., Rjasanowski, I., Wassmuth, R., Gilliam, L. K., Luo, D., and Hampe, C. S. (2005). Dynamic changes of GAD65 autoantibody epitope specificities in individuals at risk of developing type 1 diabetes. *Diabetologia* **48**, 922.

Schmidli, R. S., Colman, P. G., Bonifacio, E., Bottazzo, G. F., and Harrison, L. C. (1994). High level of concordance between assays for glutamic acid decarboxylase antibodies. The First International Glutamic Acid Decarboxylase Antibody Workshop. *Diabetes* **43**, 1005.

Schmidt, K. D., Valeri, C., and Leslie, R. D. (2005). Autoantibodies in Type 1 diabetes. *Clin. Chim. Acta* **354**, 35.

Schwartz, H. L., Chandonia, J. M., Kash, S. F., Kanaani, J., Tunnell, E., Domingo, A., Cohen, F. E., Banga, J. P., Madec, A. M., Richter, W., and Baekkeskov, S. (1999). High-resolution autoreactive epitope mapping and structural modeling of the 65 kDa form of human glutamic acid decarboxylase. *J. Mol. Biol.* **287**, 983.

Shlomchik, M. J. (2008). Sites and stages of autoreactive B cell activation and regulation. *Immunity* **28**, 18.

Shoenfeld, Y. (2004). The idiotypic network in autoimmunity: Antibodies that bind antibodies that bind antibodies. *Nat. Med.* **10**, 17.

Skarsvik, S., Tiittanen, M., Lindstrom, A., Casas, R., Ludvigsson, J., and Vaarala, O. (2004). Poor *in vitro* maturation and pro-inflammatory cytokine response of dendritic cells in children at genetic risk of type 1 diabetes. *Scand. J. Immunol.* **60**, 647.

Solimena, M., Dirkx, R., Jr, Radzynski, M., Mundigl, O., and De Camilli, P. (1994). A signal located within amino acids 1-27 of GAD65 is required for its targeting to the Golgi complex region. *J. Cell Biol.* **126**, 331.

Steed, J., Gilliam, L. K., Harris, R. A., Lernmark, A., and Hampe, C. S. (2008). Antigen presentation of detergent-free glutamate decarboxylase (GAD65) is affected by human serum albumin as carrier protein. *J. Immunol. Methods* **334**, 114.

Sundkvist, G., Hagopian, W. A., Landin-Olsson, M., Lernmark, A., Ohlsson, L., Ericsson, C., and Ahlmen, J. (1994). Islet cell antibodies, but not glutamic acid decarboxylase

antibodies, are decreased by plasmapheresis in patients with newly diagnosed insulin-dependent diabetes mellitus. *J. Clin. Endocrinol. Metab.* **78,** 1159.
Syren, K., Lindsay, L., Stoehrer, B., Jury, K., Luhder, F., Baekkeskov, S., and Richter, W. (1996). Islet cell antibodies, but not glutamic acid decarboxylase antibodies, are decreased by plasmapheresis in patients with newly diagnosed insulin-dependent diabetes mellitus. *J. Immunol.* **157,** 5208.
Takahashi, K., Honeyman, M. C., and Harrison, L. C. (1998). Impaired yield, phenotype, and function of monocyte-derived dendritic cells in humans at risk for insulin-dependent diabetes. *J. Immunol.* **161,** 2629.
TEDDY (2007). The Environmental Determinants of Diabetes in the Young (TEDDY) study: Study design. *Pediatr. Diabetes* **8,** 286.
Tian, J., Atkinson, M. A., Clare-Salzler, M., Herschenfeld, A., Forsthuber, T., Lehmann, P. V., and Kaufman, D. L. (1996a). Nasal administration of glutamate decarboxylase (GAD65) peptides induces Th2 responses and prevents murine insulin-dependent diabetes. *J. Exp. Med.* **183,** 1561.
Tian, J., Clare-Salzler, M., Herschenfeld, A., Middleton, B., Newman, D., Mueller, R., Arita, S., Evans, C., Atkinson, M. A., Mullen, Y., Sarvetnick, N., Tobin, A. J., et al. (1996b). Modulating autoimmune responses to GAD inhibits disease progression and prolongs islet graft survival in diabetes-prone mice. *Nat. Med.* **2,** 1348.
Tisch, R., Yang, X. D., Singer, S. M., Liblau, R. S., Fugger, L., and McDevitt, H. O. (1993). Immune response to glutamic acid decarboxylase correlates with insulitis in non-obese diabetic mice. *Nature* **366,** 72.
Toh, B. H., and Alderuccio, F. (2004). Pernicious anaemia. *Autoimmunity* **37,** 357.
Torn, C., Mueller, P. W., Schlosser, M., Bonifacio, E., and Bingley, P. J. (2008). Diabetes Antibody Standardization Program: Evaluation of assays for autoantibodies to glutamic acid decarboxylase and islet antigen-2. *Diabetologia* **51,** 846.
Tree, T. I., Duinkerken, G., Willemen, S., de Vries, R. R., and Roep, B. O. (2004). HLA-DQ-regulated T-cell responses to islet cell autoantigens insulin and GAD65. *Diabetes* **53,** 1692.
Tremble, J., Morgenthaler, N. G., Vlug, A., Powers, A. C., Christie, M. R., Scherbaum, W. A., and Banga, J. P. (1997). Human B cells secreting immunoglobulin G to glutamic acid decarboxylase-65 from a nondiabetic patient with multiple autoantibodies and Graves' disease: A comparison with those present in type 1 diabetes. *J. Clin. Endocrinol. Metab.* **82,** 2664.
Tuomi, T., Bjorses, P., Falorni, A., Partanen, J., Perheentupa, J., Lernmark, A., and Miettinen, A. (1996). Antibodies to glutamic acid decarboxylase and insulin-dependent diabetes in patients with autoimmune polyendocrine syndrome type I. *J. Clin. Endocrinol. Metab.* **81,** 1488.
Uibo, R., Sullivan, E. P., Uibo, O., Lernmark, A., Salur, L., Kivik, T., and Mandel, M. (2001). Comparison of the prevalence of glutamic acid decarboxylase (GAD65) and gliadin antibodies (AGA) in a randomly selected adult estonian population. *Horm. Metab. Res.* **33,** 564.
Uno, S., Imagawa, A., Okita, K., Sayama, K., Moriwaki, M., Iwahashi, H., Yamagata, K., Tamura, S., Matsuzawa, Y., Hanafusa, T., Miyagawa, J., and Shimomura, I. (2007). Macrophages and dendritic cells infiltrating islets with or without beta cells produce tumour necrosis factor-alpha in patients with recent-onset type 1 diabetes. *Diabetologia* **50,** 596.
van der Werf, N., Kroese, F. G., Rozing, J., and Hillebrands, J. L. (2007). Viral infections as potential triggers of type 1 diabetes. *Diabetes Metab. Res. Rev.* **23,** 169.
Vandewalle, C. L., Falorni, A., Svanholm, S., Lernmark, A., Pipeleers, D. G., and Gorus, F. K. (1995). High diagnostic sensitivity of glutamate decarboxylase autoantibodies in insulin-dependent diabetes mellitus with clinical onset between age 20 and 40 years. The Belgian Diabetes Registry. *J. Clin. Endocrinol. Metab.* **80,** 846.

Varela-Calvino, R., and Peakman, M. (2003). Enteroviruses and type 1 diabetes. *Diabetes Metab. Res. Rev.* **19,** 431.

Verge, C. F., Stenger, D., Bonifacio, E., Colman, P. G., Pilcher, C., Bingley, P. J., and Eisenbarth, G. S. (1998). Combined use of autoantibodies (IA-2 autoantibody, GAD autoantibody, insulin autoantibody, cytoplasmic islet cell antibodies) in type 1 diabetes: Combinatorial Islet Autoantibody Workshop. *Diabetes* **47,** 1857.

Viglietta, V., Kent, S. C., Orban, T., and Hafler, D. A. (2002). GAD65-reactive T cells are activated in patients with autoimmune type 1a diabetes. *J. Clin. Invest.* **109,** 895.

Villalba, A., Iacono, R. F., Valdez, S. N., and Poskus, E. (2008). Detection and immunochemical characterization of glutamic acid decarboxylase autoantibodies in patients with autoimmune diabetes mellitus. *Autoimmunity* **41,** 143.

Vuckovic, S., Withers, G., Harris, M., Khalil, D., Gardiner, D., Flesch, I., Tepes, S., Greer, R., Cowley, D., Cotterill, A., and Hart, D. N. (2007). Decreased blood dendritic cell counts in type 1 diabetic children. *Clin. Immunol.* **123,** 281.

Wahlberg, J., Vaarala, O., and Ludvigsson, J. (2006). Dietary risk factors for the emergence of type 1 diabetes-related autoantibodies in 2 1/2 year-old Swedish children. *Br. J. Nutr.* **95,** 603.

Waldrop, M. A., Suckow, A. T., Marcovina, S. M., and Chessler, S. D. (2007). Release of glutamate decarboxylase-65 into the circulation by injured pancreatic islet beta-cells. *Endocrinology* **148,** 4572.

Wardemann, H., and Nussenzweig, M. C. (2007). B-cell self-tolerance in humans. *Adv. Immunol.* **95,** 83.

Wasserfall, C. H., and Atkinson, M. A. (2006). Autoantibody markers for the diagnosis and prediction of type 1 diabetes. *Autoimmun. Rev.* **5,** 424.

Weetman, A. P. (2004). Autoimmune thyroid disease. *Autoimmunity* **37,** 337.

Wicker, L. S., Chen, S. L., Nepom, G. T., Elliott, J. F., Freed, D. C., Bansal, A., Zheng, S., Herman, A., Lernmark, A., Zaller, D. M., Peterson, L. B., Rothbard, J. B., *et al.* (1996). Naturally processed T cell epitopes from human glutamic acid decarboxylase identified using mice transgenic for the type 1 diabetes-associated human MHC class II allele, DRB1*0401. *J. Clin. Invest.* **98,** 2597.

CHAPTER 4

CD8+ T Cells in Type 1 Diabetes

Sue Tsai, Afshin Shameli, and **Pere Santamaria**

Contents		
	1. Introduction	80
	2. MHC Class I and T1D	81
	3. Autoreactive CD8+ T Cells in T1D	83
	4. Antigens for Diabetogenic CD8+ T Cells in Humans	86
	4.1. GAD65	91
	4.2. GFAP	91
	4.3. IA-2	92
	4.4. IGRP	92
	4.5. Insulin	93
	5. The Relative Contribution of the β-Cell-Specific CD8+ Response to T1D	95
	6. Development and Activation of Diabetogenic CD8+ T Cells	96
	6.1. MHC class I-restricted TCR-transgenic models of T1D	97
	6.2. Activation of diabetogenic CD8+ T cells	98
	6.3. The requirement for CD4+ T cell help	99
	6.4. Inflammatory signals augment diabetogenic CD8+ T cell activation	100
	6.5. Defective immune tolerance fuels the activation of diabetogenic CD8+ T cells	102
	6.6. Cross-tolerance of diabetogenic CD8+ T cells	103
	7. Recruitment of Diabetogenic CD8+ T Cells to Islets	104
	8. Mechanisms of β-Cell Cytotoxicity in T1D	105
	8.1. Fas versus granule exocytosis	105
	8.2. TNF-α, TNFR, IFN-γ, and IL-1	107

Julia McFarlane Diabetes Research Centre (JMDRC), Department of Microbiology and Infectious Diseases, and Calvin, Phoebe and Joan Snyder Institute of Infection, Immunity and Inflammation, Faculty of Medicine, University of Calgary, Calgary, Alberta, Canada

Advances in Immunology, Volume 100 © 2008 Elsevier Inc.
ISSN 0065-2776, DOI: 10.1016/S0065-2776(08)00804-3 All rights reserved.

9. Induction of Immunologic Tolerance in Diabetogenic
   CD8+ T Cells   108
10. Concluding Remarks   111
Acknowledgments   112
References   112

Abstract   Type 1 diabetes (T1D), an autoimmune disease once thought to be mediated exclusively by β cell-specific CD4+ T cells, is now recognized as one in which autoreactive CD8+ T cells play a fundamental role. In the nonobese diabetic (NOD) mouse model, CD8+ T effector cells take centre stage in the destruction of pancreatic β cells and contribute to sustaining islet inflammation. Recent investigations have elucidated the mechanisms underlying the activation, homing, and beta cell destructive properties of this type of cells. Another important area is the development and testing of novel preemptive or therapeutic "vaccines" that, by targeting effector and/or regulatory autoreactive CD8+ T cell specificities may be able to induce immunological tolerance to β cells. In humans, our understanding of the role of CD8+ T cells in T1D is also growing, through genetic linkage analyses, as well as epitope identification and characterization of disease-relevant CD8+ T cell responses in patient blood samples. The following review discusses these important advances and how they can converge towards the goal of developing an antigen-specific immunotherapy for T1D.

1. INTRODUCTION

Immune tolerance to self is maintained through a series of checkpoints that operate in both the thymus and periphery (Walker and Abbas, 2002). In the thymus, developing T cells bearing T cell receptors that engage self peptide/major histocompatibility complex (pMHC) molecules with high affinity are deleted by apoptosis in a process known as negative selection. In contrast, self-reactive T cells that recognize self pMHC to some degree but with inadequate affinity, or those whose cognate autoantigens are expressed at very low levels (or not at all) in the thymus, are spared death and thus populate the peripheral immune system (Bouneaud *et al.*, 2000). Indeed, self-reactive T cells are found not only in patients suffering from autoimmune disorders, but also in healthy individuals. Hence, whether an individual will develop autoimmunity or not depends on the balance between these potentially pathogenic self-reactive immune cell types and the regulatory mechanisms that keep them in check. These regulatory mechanisms may be T cell intrinsic, such as the induction of anergy and activation-induced cell death upon interaction with tolerogenic antigen-presenting cells (APCs) (Steinman *et al.*, 2003), or may be mediated by immune cell types with regulatory activity (e.g., regulatory T cells, Tregs).

In Type 1 Diabetes (T1D), there is a loss of tolerance to pancreatic β cells that results in a CD4+ and CD8+ T cell-dependent autoimmune process that culminates in complete destruction of insulin-producing β cells, leading to insulin deficiency and dysregulated glucose metabolism. How to restore self-tolerance to β cells remains an unresolved problem of fundamental importance in diabetes research. Over the years, studies using the non-obese diabetic (NOD) mouse model have provided important insights into the immunology of T1D and have demonstrated the feasibility of preventing or curing T1D in an antigen-specific manner by specifically manipulating disease-relevant autoreactive T cell populations without compromising the immune system at large. However, the translation of these findings to the clinic remains a promise rather than a reality for numerous reasons, and awaits both a better understanding of the disease process and the development of novel interventional strategies that can readily be translated into patient health-care. Here, we discuss recent advancements made in this field, focusing on different aspects of the biology of diabetogenic CD8+ T cell populations that might serve as targets for therapeutic intervention: their development, activation and recruitment, antigenic specificities, pathogenicity, and β cell-killing properties and mechanisms.

2. MHC CLASS I AND T1D

T1D in the NOD mouse shares several genetic and immunological features with the human condition. These include, for instance, the phenomenon of polygenic inheritance in which numerous non-MHC-linked genes modulate dominant susceptibility afforded by MHC-linked loci, the slow progression of β cell destruction towards overt clinical disease, and the involvement of a highly polyclonal autoimmune response that targets numerous epitopes in many β cell antigens and involves both CD4+ and CD8+ T cells, among other cell types of the immune system such as B lymphocytes, dendritic cells (DCs), and macrophages (Anderson and Bluestone, 2005; Wicker *et al.*, 2005). In both mouse and humans, T1D susceptibility is determined by the constellation of alleles at multiple genetic loci that afford either susceptibility or resistance to the disease (referred to as *Idd* and *IDDM*, respectively) (Todd and Wicker, 2001). Of the loci identified to date, the MHC region is the primary genetic determinant of diabetes, as is also the case in other autoimmune disorders.

The NOD mouse expresses a unique H-2^{g7} haplotype (H-2K^d, H-2D^b, I-A^{g7}, and I-E^{null}), that encodes common class I allelic variants (K^d and D^b), an I-Eβ chain that cannot be expressed on the surface of APCs because this haplotype bears a defective Eα gene, and a unique MHC class II molecule (I-A^{g7}). The NOD I-Aβ chain carries histidine and serine residues at positions 56 and 57 as opposed to the proline and aspartic acid residues commonly in I-Aβ chains encoded by haplotypes found in diabetes-resistant strains of mice (Acha-Orbea and McDevitt, 1987), an observation that is strikingly recapitulated in the diabetes-promoting HLA-DQB1 alleles in humans (Todd et al., 1987). The NOD mouse requires both copies of the I-A^{g7} gene to develop diabetes, as the introduction of a single copy of a diabetes-protective MHC class II gene, I-E or I-A, into the NOD genetic background efficiently abrogates diabetes. How these MHC class II molecules afford diabetes resistance remains unclear. However, several possible mechanisms have been proposed. These include promoting negative selection of diabetogenic CD4+ and CD8+ T cell clonotypes in the thymus (Schmidt et al., 1997, 1999; Serreze et al., 2004; Thiessen et al., 2002), induction of peripheral tolerance of these T cells via deletion or functional inactivation (anergy), and induction of regulatory T cells (Tregs) able to keep these autoreactive T cell specificities in check. The observation that certain diabetogenic MHC class I-restricted CD8+ T cells might recognize protective MHC class II allelic variants in the thymus is an interesting one (Serreze et al., 2004) and suggests that MHC promiscuity might be a common feature of pathogenic autoreactive CD8+ T cells, as we had previously proposed for pathogenic autoreactive CD4+ T cells (Schmidt et al., 1997).

The realization, many years ago, that diabetes susceptibility/resistance was strongly influenced by MHC class II polymorphism fostered the belief that CD4+ T cells might be the major effectors of β cell destruction in T1D (Tisch and McDevitt, 1996). It is now clear, however, that although CD4+ T cells are required for diabetes progression, T1D development requires important contributions from MHC class I-restricted autoreactive CD8+ T cells at different stages of the disease. Certain MHC class I genes have been linked to T1D susceptibility when expressed in conjunction with certain MHC class II susceptibility alleles in humans (Langholz et al., 1995; Noble et al., 2002; Robles et al., 2002; Tait et al., 2003). Recently, Nejentsev et al. have demonstrated a direct contribution of HLA-A and HLA-B to T1D, independently of the MHC class II susceptibility loci (Nejentsev et al., 2007). As the main function of MHC class I molecules is the presentation of antigenic peptides to CD8+ T cells, these MHC class I molecules likely contribute to diabetogenesis by shaping the repertoire of peripheral autoreactive CD8+ T cells. Thus, when introduced into a susceptible background, such as in the NOD mouse, the expression of an HLA-A2 transgene accelerated diabetes onset (Marron et al., 2002). In the NOD mouse, the link between MHC class I genes and diabetes susceptibility is less clear, since the NOD MHC class I molecules, H-2K^d and H-2D^b, are common and found also in diabetes-resistant mouse strains. Nevertheless, certain observations do suggest that the NOD MHC class I genes interact with other *Idd* loci to provide an environment that is permissive to the development of pathogenic CD8+ T cells (reviewed in DiLorenzo and Serreze, 2005).

3. AUTOREACTIVE CD8+ T CELLS IN T1D

Some of the earliest direct evidence for the involvement of autoreactive CD8+ T cells in T1D came from histological studies of pancreata from diabetic patients. These studies documented significant islet CD8+ T cell infiltration in recently diagnosed diabetic patients (Hanninen et al., 1992; Itoh et al., 1993), as well as in pancreas graft biopsies from diabetic recipients treated with pancreas isografts or HLA-identical allografts (Itoh et al., 1993; Santamaria et al., 1992a,b; Somoza et al., 1994). Prompted by these observations, we and others set out to study the CD8+ T cell compartment in the NOD mouse. These studies have established a major contribution of autoreactive CD8+ T cells in the pathogenesis of T1D.

The independent contribution of CD4+ and CD8+ T cells to diabetogenesis was first assessed in a series of adoptive transfer studies that compared the ability of individual NOD T cell subsets to transfer diabetes into severe-combined immunodeficient NOD (NOD.*scid*) hosts or NOD neonates (Bendelac et al., 1987; Christianson et al., 1993; DiLorenzo et al., 1998; Miller et al., 1988; Yagi et al., 1992). These studies provided compelling support for the idea that diabetogenesis requires both T cell subsets. For example, co-transfer of splenic CD8+ and CD4+ T cells from prediabetic NOD females into NOD.*scid* mice induced diabetes more efficiently than the transfer of CD4+ T cells alone (Christianson et al., 1993). In agreement with the results of these studies, NOD islet-derived, β cell-specific cytotoxic T lymphocyte (CTL) lines and clones were found to be capable of transferring diabetes, particularly when given with unstimulated polyclonal CD4+ T cells (Nagata et al., 1994; Utsugi et al., 1996). It was subsequently shown that NOD mice either lacking the β2-microglobulin (β2m) gene or expressing a functionally disrupted CD8α chain developed neither diabetes nor insulitis (DiLorenzo et al., 1998; Serreze et al., 1994, 1997; Sumida et al., 1994; Wang et al., 1996). Most notably, NOD mice lacking MHC class I exclusively on β cells or APCs failed to develop diabetes, supporting a major contribution of CD8–APC and CD8–β cell interactions to the progression of diabetes (Hamilton-Williams et al., 2003; de Jersey et al., 2007; Yamanouchi et al., 2003).

CD8+ T cells are amongst the first to infiltrate pancreatic islets. As early as at 3 weeks of age, a population of CD8+ T cells recognizing an insulin-derived epitope (mIns B15–23) was reported to infiltrate the islets of NOD mice (Wong et al., 1999). The size of this population declines with age and is replaced by highly diabetogenic CD8+ T cell specificities, including one that uses a highly conserved TCRα rearrangement devoid of CDR3α variability (Vα17-Jα42) (DiLorenzo et al., 1998; Verdaguer et al., 1996, 1997). This population of CD8+ T cells recognizes a H-2K^d-restricted epitope consisting of residues 206–214 of the β cell antigen Islet-associated Glucose-6-phosphatase catalytic subunit-Related Protein (IGRP), as well as several peptide mimics of IGRP$_{206-214}$, such as NRP-V7 (a super-agonist), NRP-A7 (an agonist), and NRP-I4 (a partial agonist) (Lieberman et al., 2003; Santamaria et al., 1995; Verdaguer et al., 1996, 1997).

The role of IGRP$_{206-214}$-reactive CD8+ T cells in NOD diabetes is well established. IGRP$_{206-214}$-reactive CD8+ T cells are highly diabetogenic (Verdaguer et al., 1997) and prevalent in both the circulation and pancreatic islets, accounting for up to 0.5% and 40% of the corresponding CD8+ T cell populations, respectively (Trudeau et al., 2003). As the disease progresses from benign insulitis to overt clinical disease, IGRP$_{206-214}$-specific CD8+ T cells undergo a process of avidity maturation, whereby high-avidity CD8+ T cell clones gradually out-compete and replace their lower-avidity counterparts, resulting in an overall increase in the binding avidity of the T cell pool for its cognate pMHC ligand (Amrani et al., 2000a). It is likely that such a phenomenon is not unique to the IGRP$_{206-214}$-reactive CD8+ T cell population, but rather shared by other pathogenic autoreactive T cell specificities, although this remains to be demonstrated.

Another unique feature of the IGRP$_{206-214}$-reactive CD8+ T cell pool is that it undergoes cyclic expansion and contraction in the circulation prior to the onset of hyperglycemia. Notably, increases in the peripheral frequency of this CD8+ T cell population can be used to predict which mice within the colony will develop diabetes and which ones will not, with a high degree of certainty (Trudeau et al., 2003). It is also noteworthy that IGRP$_{206-214}$-reactive CD8+ T cells are the predominant CD8+ T cell subset infiltrating NOD islet isografts, likely contributing to the early stages of islet graft rejection in this model (Wong et al., 2006).

Over the last few years, several studies have reported that the diabetogenic CD8+ T cell response in NOD mice involves recognition of other epitopes within IGRP (Han et al., 2005a), as well as epitopes in β cell antigens other than IGRP and insulin (Table 4.1). In addition, CD8+ T cell clones recognizing some of these epitopes have been shown to have diabetogenic or insulitic potential in adoptive transfer studies (Bowie et al., 1999; Busick et al., 2007; Quinn et al., 2001; Severe et al., 2007; Wong et al., 1996, 1999). Taken together, the above studies in NOD mice have demonstrated that autoreactive CD8+ T cells play a pivotal

TABLE 4.1 $H-2K^d$ and $H-2D^b$-restricted epitopes for T1D

Antigen	Epitope	Amino acid sequence	MHC Class I	Type of Response	Source of T cells /clone	TCR-Tg	Reference	Comments
DMK	138-146	FQDENYLYL	$H-2D^b$	Spontaneous	islet / AI4	NOD.AI4$\alpha\beta$	Lieberman et al., 2004	DEC-205-targetted superagonist MimA2 tolerized AI4 CD8+ T cells (Mukhopadhaya et al. 2008)
	88-98	NYAFLHATDLL	$H-2K^d$	peptide-imm	spleen		Busick et al., 2007	
	118-128	LLQYVVKSFDR	$H-2K^d$	peptide-imm	spleen		Busick et al., 2007	
GAD65	206-214	TYEIAPVFV	$H-2K^d$	Spontaneous	spleen		Quinn et al., 2001	
	268-278	AAVPRLIAPTS	$H-2K^d$	peptide-imm	spleen		Busick et al., 2007	
	507-516	WFVPPSLRTL	$H-2K^d$	peptide-imm	spleen		Videbaek et al., 2003	
	546-554	SYQPLGDKV	$H-2K^d$	Spontaneous	spleen		Quinn et al., 2001; Busick et al., 2007	
GAD67	515-524	WYIPQSLRGV	$H-2K^d$	peptide-imm	spleen		Bowie, Tite and Cooke, 1999	
GFAP	79-87	KALAAELNQ	$H-2K^d$	Spontaneous	Spleen		Tsui et al., 2008	Peptide/IFA immunotherapy induced diabetes protection in NOD mice.
	253-261	SRNAELLRQ	$H-2K^d$	Spontaneous	Spleen		Tsui et al., 2008	
IGRP [a]	21-29	TYYGFLNFM	$H-2K^d$	Spontaneous	islet		Han et al., 2005a	
	206-214	VYLKTNVFL	$H-2K^d$	Spontaneous	islet / NY8.3	8.3-NOD 17.5-NOD 17.6-NOD	Lieberman et al., 2004; Han et al., 2005b	APL therapy with agonistic mimotope NRP-A7 induced diabetes protection in NOD mice at specific doses.
	225-233	LRLFGIDLL	$H-2D^b$	Spontaneous	islet		Han et al., 2005a	
	241-249	KWCANPDWI	$H-2D^b$	Spontaneous	islet		Han et al., 2005a	
	324-332	SFCKSASIP	$H-2K^d$	Spontaneous	islet		Han et al., 2005a	
Insulin 1/2	B15-23	LYLVCGERG	$H-2K^d$	Spontaneous	islet / G9C8		Wong et al., 1999; Wong et al., 1996	
Insulin 2	B25-C2	FYTPMSRREV	$H-2K^d$	peptide-imm	spleen		Martinez et al., 2003	Peptide/IFA immunotherapy induced diabetes protection in NOD mice.

[a] The listed epitopes include only those recognized by prevalent clonotypes in the islets (i.e. detectable by $H-2K^d$ or $H-2D^b$ tetramer staining). Islet CTL IFN-γ responses to other IGPR epitopes have also been detected (Han et al., 2005a).

role in T1D and have prompted further research efforts to explore the contribution of this T cell subset in human T1D. In particular, the past few years have seen considerable advancements in epitope discovery.

4. ANTIGENS FOR DIABETOGENIC CD8+ T CELLS IN HUMANS

Epitope discovery is an important area of research in T1D: CD8+ T cells are potential markers of disease onset or progression and are also considered as emerging targets for prospective antigen-specific therapeutic interventions. MHC class I-restricted epitope identification in the NOD mouse has led to innovative applications such as the design of preventative peptide vaccines and combinatorial therapies (Bresson *et al.*, 2006; Han *et al.*, 2005a). In humans, efforts to identify disease-relevant CD8+ T cell specificities had been hindered by a number of factors, including limited access to patient samples, scarcity of these cells in the peripheral blood, and other technical limitations. Recently, an ELISpot-based approach has been developed to assay for CD8+ T cell responses against β cell antigens, using peripheral blood mononuclear cells (PBMCs) of diabetic patients and at risk individuals. This approach, when combined with the use of HLA-A2-transgenic NOD mice, has become a powerful tool for identifying novel HLA-A2-restricted epitopes in T1D (Serreze *et al.*, 2007). HLA-A2 is a frequent allele in Caucasian populations, and would afford wide population coverage to any epitope-based immunointerventional strategy based on it. The recent identification of T1D-linked HLA-A and HLA-B alleles (e.g., HLA-A24 and HLA-B39) (Nejentsev *et al.*, 2007), warrants the search for autoantigenic specificities presented by these class I molecules in the context of T1D.

Looking at recent reports in this area, a common starting point has been the selection of candidate autoantigenic peptides based on their predicted MHC class I-binding motifs, using algorithms such as BIMAS (which bases its prediction on the strength of interaction between peptide and MHC) (Parker *et al.*, 1994) or SYFPEITHI (an algorithm that is trained on actual epitope data) (Rammensee *et al.*, 1999). The actual strength of binding between the peptides and MHC class I molecules can also be confirmed experimentally via a MHC-stabilization assay (Nijman *et al.*, 1993). On this point, it should be noted that the predicted MHC binding affinity may differ from the measured value, as reported by some researchers (Standifer *et al.*, 2006). More importantly, the immunogenicity of a given pMHC complex does not necessarily correlate with the peptide's MHC class I-binding affinity. One group reported an inverse relationship between the pMHC class I binding affinity and the number of responses observed in diabetic patients (Ouyang *et al.*, 2006), and another observed CD8+ responses against a poor-binding peptide (Jarchum et al., 2008). In another study by Baker and colleagues, it was noted that insulin-specific CD8+ T cell responses are often elicited to intermediate to lower-affinity peptides (for MHC) (Baker et al., 2008). These observations are not surprising, as low-affinity peptides may not adequately mediate the negative selection of cognate T cell specificities in the thymus, leading to their escape into the periphery. The screening of a complete and unbiased library of proinsulin-derived peptides provided an appreciation for the breadth of responses harbored by diabetic patients, as well as additional clues for the identification of novel epitopes that were perhaps overlooked in studies that biased epitope selection towards high-affinity binders (Baker et al., 2008).

In addition to the requirement for MHC class I binding and presentation, whether the peptide epitopes arise naturally (i.e., through antigen-processing by the APC's intracellular machinery) is another factor to consider. In the process of generating a MHC class I-restricted peptide, the proteasome cleaves at the carboxy terminus of the peptide, leaving it to be trimmed at the N-terminus by other cytoplasmic or ER-resident peptidases (Shastri et al., 2002). Algorithms that predict proteasome cleavage sites on the C-terminus of the candidate peptides are available (Kesmir et al., 2002). Alternatively, smaller antigenic polypeptides can be subjected to *in vitro* digestion with proteasome extracts (Hassainya et al., 2005; Toma et al., 2005), the products from which are then identified and used to screen for CD8+ T cell responses.

From here, the selected peptides were tested against PBMCs of HLA-class I-matched individuals of different disease statuses, including at risk individuals that harbored β cell-specific autoantibodies, recent-onset and long-standing diabetic patients, as well as patients that had received islet transplants and subsequently developed recurrent diabetes, in comparison to healthy individuals. In other studies, the peptides were selected based on their ability to elicit recall responses in antigen-immunized HLA-A2-transgenic NOD mice (NOD.β2m^{null}.HHD), or their recognition by spontaneously arising islet-infiltrating CD8+ T cells; these findings were then subjected to validation in human subjects.

As a result, several novel T1D-associated HLA class I-restricted epitopes have been identified and include epitopes from 65 kDa glutamic acid decarboxylase (GAD) (GAD65), prepro-islet amyloid polypeptide (ppIAPP), Insulinoma-associated antigen-2 (IA-2), islet-associated glucose 6-phosphatase catalytic subunit-related protein (IGRP), and glial fibrillary acidic protein (GFAP), as well as several epitopes in the proinsulin polypeptide. Here, we attempt to summarize these current findings into a comprehensive table (Table 4.2), with a brief description of each antigen category below.

TABLE 4.2 HLA CLASS I-restricted epitopes in T1D

Antigen	Epitope	HLA	Amino Acid Sequence	Responses in HHD mice	Type of response	Responses in human PBMC	HLA class I binding[e]	Proteasomal digestion[e]	References
GAD65	114-123	A2	VMNILLQYVV	Yes	plasmid-imm	Yes (R, A)[c]	Yes		Blancou et al. 2007; Mallone et al. 2007; Panina-Bordignon et al. 1995
	536-545	A2	RMMEYGTTMV	Yes	diabetic/plasmid-imm	Yes (R)[c]	Yes		Blancou et al. 2007; Enee et al. 2008
GFAP	143-151	A2	NLAQTDLATV			Yes (R, I, A)[cg]	Yes		Standifer et al. 2006
	192-200	A2	SLEEIRFL			Yes (R, A)[cg]	Yes		Standifer et al. 2006
	214-222	A2	QLARQQVHV			Yes (R, I, A)[cg]	Yes		Standifer et al. 2006
IA-2	172-180	A2	SLSPLQAEL			Yes (R)[c]	Yes		Ouyang et al. 2006
	482-490	A2	SLAAGVKLL			Yes (R)[c]	Yes		Ouyang et al. 2006
	797-805	A2	MVWESGCTV			Yes (A)[c]	Yes		Takahashi et al. 2001
	805-813	A2	VIVMLTPLV	Yes	plasmid-imm	Yes (R)[c]	Yes		Blancou et al. 2007
	830-839	A2	SLYHVEVNL	Yes	diabetic/plasmid-imm	Yes (R)[c]	Yes		Blancou et al. 2007; Enee et al. 2008
ppIAPP	5-13	A2	KLQVFLIVL			Yes (R, L, A, P)[a]	Yes		Jarchum et al. 2008; Ouyang et al. 2006; Panagiotopoulos et al. 2003; Standifer et al. 2006
	9-17	A2	FLIVLSVAL			Yes (R, I, A)[ag]	Yes		Ouyang et al. 2006; Standifer et al. 2006
IGRP	152-160	A2	FLWSVFMLI			Yes (R, I, A)[cg]	Yes		Ouyang et al. 2006; Standifer et al. 2006
	211-219	A2	NLFLFLFAV			Yes (P)[c]	undetectable		Jarchum et al. 2008
	215-223	A2	FLFAVGFYL			Yes (R, A, P)[ag]	Yes		Jarchum et al. 2008; Ouyang et al. 2006; Standifer et al. 2006
	222-230	A2	YLLLRVLNI			Yes (P)[a]	Yes		Jarchum et al. 2008
	228-236	A2	LNIDLLWSV	Yes	prediabetic	Yes (R)[c]	Yes		Takaki et al. 2006; Mallone et al. 2007
	265-273	A2	VLFGLGFAI	Yes	prediabetic	Yes (R, P)[c]	Yes		Takaki et al. 2006; Unger et al. 2007; Jarchum et al. 2008
	293-301	A2	RLLCALTSL			Yes (R, A)[cg]	Yes		Ouyang et al. 2006; Standifer et al. 2006

	L2-10	A2	ALWMRLLPL	Yes		Yes (R)[c]	Yes	Ouyang et al. 2006; Mallone et al. 2007; Jarchum et al. 2007
	L3-11	A2	LWMRLLPLL	Yes[f]	peptide-imm		undetectable	Jarchum et al. 2007
	B5-14	A2	HLCGSHLVEA	Yes	prediabetic / diabetic		Yes	Jarchum et al. 2007
	B10-18	A2	HLVEALYLV	Yes	diabetic / peptide-imm	Yes (R, L, T)[a]	Yes	Toma et al. 2005; Hassainya et al. 2005; Pinkse et al. 2005; Enee et al. 2008
	B-14-22	A3	ALYLVCGER			Yes (R)[c]	Yes	Toma et al. 2005
	B14-22	A11	ALYLVCGER			Yes (R, L)[c]	Yes	Toma et al. 2005
	B15-24	A24	LYLVCGERGF			Yes (R, L)[c]	Yes	Kimura et al. 2001; Toma et al. 2005
	B17-26	A1	LVCGERGFFY			Yes (R, L)[a]	Yes	Toma et al. 2005
	B17-26	A3	LVCGERGFFY			Yes (R, L)[c]	Yes	Toma et al. 2005
	B18-27	A1	VCGERGFFYT			Yes (R, L)[c]	Yes	Toma et al. 2005
proinsulin	B18-27	A2	VCGERGFFYT	Yes	peptide-imm	Yes (R, L)[a]	Yes	Toma et al. 2005; Hassainya et al. 2005; Mallone et al. 2007
	B18-27	B8	VCGERGFFYT			Yes (R, L)[a]	Yes	Toma et al. 2005
	B18-27	B18	VCGERGFFYT			Yes (R, L)[d]	Yes	Toma et al. 2005
	B20-27	A1	GERGFFYT			Yes (R, L)[c]	Yes (Low)	Toma et al. 2005
	B20-27	B8	GERGFFYT			Yes (R, L)[a]	Yes (Low)	Toma et al. 2005
	B21-29	A3	ERGFFYTPK			Yes (R, L)[c]	Yes	Toma et al. 2005
	B25-C1	B8	FYTPKTRRE			Yes (R, L)[a]	Yes	Toma et al. 2005
	B27-C5	B8	TPKTRREAEDL			Yes (R)[b]	Yes (Low)	Toma et al. 2005
	C20-28	A2	SLQPLALEG	Yes	peptide-imm		Yes	Hassainya et al. 2005
	C25-33	A2	ALEGSLQKR	Yes	peptide-imm		Yes	Hassainya et al. 2005
	C29-A5	A2	SLQKRGIVEQ	Yes	peptide-imm		Yes	Hassainya et al. 2005

Antigen	Epitope	HLA	Amino Acid Sequence	Responses in HHD mice	Type of response	Responses in human PBMC	HLA class I binding[e]	Proteasomal digestion[e]	References
	A1-10	A2	GIVEQCCTSI	Yes	peptide-imm		Yes	Yes	Hassainya et al. 2005
	A2-10	A2	IVEQCCTSI	Yes	prediabetic / diabetic / peptide-imm		undetectable	Yes	Jarchum et al. 2007
	A12-20	A2	SLYQLENYC	Yes	diabetic / peptide-imm	Yes (R)[c]	Yes	No	Hassainya et al. 2005; Mallone et al. 2007; Enee et al. 2008

Antigens: GAD65: 65kDa Glutamic acid decarboxylase; GFAP: glial fibrillary acidic protein; IA-2: insulinoma-associated antigen 2; ppIAPP: Islet amyloid polypeptide precursor protein; IGRP: Islet-specific glucose 6-phosphatase catalytic subunit-related protein

Patient subsets: R: recent-onset diabetic defined as within 6 months of dianosis; I: intermediate duration of diabetes (6 months < diabetes duration < 5 years; L: long-standing diabetic defined as being insulin-dependent for > 5 years; A: at risk individuals ie, first-degree relatives of diabetic patients that are positive for serum Ab against IA-2, GAD65, or islet cell Ab; P: pediatric diabetic patients; T: recurrent diabetic islet transplant recipients

Statistical analysis: *a*: statiscally significant compared to control; *b*: not statistically significant compared to control; *c*: qualitative difference between control and patients, statistical significance not determined; *d*: without negative controls;

e: determined experimentally

f: reactivity towards the murine insulin 1 epitope was detected

g: positive granzyme B responses determined as well as IFNγ

4.1. GAD65

GAD catalyzes the synthesis of the neurotransmitter gamma-amino butyric acid and is found in the brain as well as in pancreatic islets. In T1D, islet GAD is a target of autoimmune attack by autoantibodies, as well as by CD4+ and CD8+ T cells. Mouse islets, predominantly β cells, express two forms of GAD with predicted molecular weights of ∼65 kDa (GAD65) and 67 kDa (GAD67). Because GAD65 was also found to be expressed in human islets (Baekkeskov et al., 1990), it captured the attention of many investigators as a major candidate autoantigen for diabetes in both NOD mice and human patients. However, since NOD mouse islet cells express much lower levels of GAD65 than human islets (Kim et al., 1993), GAD65-specific responses found in the NOD mouse may be of a different nature than those found in humans. In the NOD mouse, GAD65-specific T cell responses have been associated with both pathogenesis and regulation (Quinn et al., 2001; Tarbell et al., 2002; Tisch et al., 1993; Videbaek et al., 2003). In humans, the presence of GAD65-specific antibodies has been used as a diagnostic/predictive marker of T1D (Achenbach et al., 2008; Verge et al., 1996). HLA-A2-restricted GAD-specific CD8+ T cells are present in the peripheral blood of at risk individuals and recent-onset diabetes patients (Blancou et al., 2007; Mallone et al., 2007; Panina-Bordignon et al., 1995). These include $GAD_{536-545}$-specific responses, which were first identified in plasmid DNA-immunized NOD.β2m^{null}.HHD mice (Blancou et al., 2007), as well as responses targeting $GAD_{114-123}$ (Mallone et al., 2007; Panina-Bordignon et al., 1995).

4.2. GFAP

An increasing body of evidence suggests that T1D does not exclusively involve β cell-specific autoimmunity, and that neuronal elements are also targeted in the disease process as well (Tsui et al., 2007). Indeed, a number of T1D autoantigens are not only expressed in pancreatic β cells, but also in neuroendocrine cell types or nerve terminals (e.g., GAD65 discussed above, IA-2 below). A recent example is the GFAP, which is expressed on the peri-islet Schwann cells (pSC) that envelope pancreatic islets (Winer et al., 2003). Winer and colleagues observed spontaneous T cell autoimmunity against pSC in NOD mice, and this event was found to precede noticeable β cell loss. The role of pSC-specific CD8+ T cell responses in NOD diabetes was demonstrated in a subsequent study, which showed that transgenic restoration of β2m expression on pSCs in the diabetes-resistant NOD.β2m$^{-/-}$ background significantly accelerated adoptively transferred diabetes (Tsui et al., 2008). From this study, two H-2K^d-restricted epitopes, $GFAP_{79-87}$ and $GFAP_{253-261}$, were identified, and $GFAP_{79-87}$-specific immunotherapy afforded significant diabetes protection in NOD mice (Tsui *et al.*, 2008), indicating the contribution of pSC-specific CD8+ responses to diabetes development. In human patients, GFAP-specific T cell responses have also been detected in the peripheral blood (Winer *et al.*, 2003). CD8+ T cell-associated granzyme B and IFNγ responses were observed to target the HLA-A2-restricted epitopes $GFAP_{143-151}$, $GFAP_{192-200}$, and $GFAP_{214-222}$ in recent-onset patients, as well as at-risk individuals (Standifer *et al.*, 2006).

4.3. IA-2

Insulinoma-associated antigen-2 (IA-2) is a 106 kDa transmembrane protein localized in the dense core secretory vesicles of neuroendocrine tissues including the brain, pituitary, and pancreatic islets (Lan *et al.*, 1996). IA-2 belongs to the receptor-type protein tyrosine phosphatase family; although its catalytic activity/function is largely unknown, data from knock-out studies suggest IA-2 is involved in the regulation of insulin release and glucose tolerance (Henquin *et al.*, 2008; Saeki *et al.*, 2002). Like GAD65, the presence of IA-2-specific autoantibodies in the serum serves as a surrogate marker for T1D (Verge *et al.*, 1996). IA-2-specific CD8+ T cell responses have been documented in recent-onset diabetic patients (Blancou *et al.*, 2007; Ouyang *et al.*, 2006) and antibody-positive first-degree relatives (Takahashi *et al.*, 2001). The epitopes recognized by these CD8+ T cell specificities include $IA-2_{172-180}$, $IA-2_{482-490}$ (Ouyang *et al.*, 2006), $IA-2_{797-805}$ (Takahashi *et al.*, 2001), $IA-2_{805-813}$, and $IA-2_{830-839}$ (Blancou *et al.*, 2007).

4.4. IGRP

IGRP is an ER-resident, nine-span transmembrane protein that is expressed predominantly in β cells and to a lesser extent, in alpha cells (Arden *et al.*, 1999; Martin *et al.*, 2001). IGRP is a homolog of the glucose-6-phosphatase found in the liver, and may possess some catalytic activity (Petrolonis *et al.*, 2004). In the NOD mouse, $H-2K^d$ and $H-2D^b$-restricted IGRP-specific CD8+ T cells represent a significant fraction of the islet-infiltrating population (Lieberman *et al.*, 2003; Santamaria *et al.*, 1995; Verdaguer *et al.*, 1996, 1997). IGRP is also emerging as one of the major β cell antigens targeted by CD8+ T cells in human T1D. Several IGRP-specific, HLA-A2-restricted T cell specificities have been identified. $IGRP_{152-160}$, $IGRP_{215-223}$, and $IGRP_{293-301}$ were found to elicit both IFNγ and Granzyme B responses in recent-onset diabetic as well as antibody-positive individuals (Ouyang *et al.*, 2006; Standifer *et al.*, 2006). Takaki and colleagues identified spontaneous islet CD8+ T cell responses towards three HLA-A2-restricted, murine IGRP-derived peptides, $IGRP_{228-236}$, $IGRP_{265-273}$, and $IGRP_{337-345}$ in HLA-A2-transgenic mice (Takaki *et al.*, 2006). Of these, the $IGRP_{265-273}$

epitope was found to be identical in human and mouse IGRP, and presence of $IGRP_{265-273}$–reactive CD8+ T cells in recent-onset diabetic patients was reported in two independent investigations (Jarchum et al., 2008; Unger et al., 2007). Human $IGRP_{228-236}$ differs from the murine sequence at two amino acid residues, and HLA-A2-restricted $IGRP_{228-236}$-specific CD8+ T cell responses have also been reported in recent-onset-diabetic patients by Mallone and colleagues (Mallone et al., 2007). Furthermore, Jarchum et al identified CD8+ reactivities towards both $IGRP_{215-223}$ and $IGRP_{265-273}$, as well as two additional IGRP epitopes, $IGRP_{211-219}$ and $IGRP_{222-230}$, in pediatric patients (Jarchum et al., 2008), complementing the adult patient studies mentioned above.

4.5. Insulin

Both NOD mice and human patients mount humoral and T cell responses against insulin and proinsulin. The NOD mouse expresses two pre-proinsulin genes, insulin 1 and insulin 2, with insulin 1 expressed in the pancreas and insulin 2 in both the thymus and the pancreas. Insulin 2-deficient NOD mice develop accelerated diabetes (Thebault-Baumont et al., 2003), whereas insulin 1-deficient NOD mice develop insulin autoantibodies but are free of insulitis and diabetes (Moriyama et al., 2003). The insulin B9-23 epitope appears to be critical for diabetes initiation, as NOD mice lacking endogenous insulin (both insulin-1 and 2) but expressing an insulin transgene with a mutated B9-23 epitope are completely protected from diabetes (Nakayama et al., 2005). However, caution must be exerted when interpreting these data. Incidence studies carried out by the T1D repository (T1DR) showed that the congenic control strains used in this study do not develop diabetes at the same extent as reported by the authors, suggesting that genetic contamination originated from the speed-congenic approach may partially account for the observed phenotype (http://type1diabetes.jax.org/gqc.html).

In humans, the role of insulin-reactive CD8+ T cells in T1D pathogenesis is less certain, although multiple CD8+ T cell reactivities against insulin have been detected in diabetic patients by several investigators (Table 4.2). Independent studies reproducibly identified IFNγ ELISpot responses targeting InsL2-10 (Jarchum et al., 2007; Mallone et al., 2007; Ouyang et al., 2006), InsB10-18 (Hassainya et al., 2005; Pinkse et al., 2005; Toma et al., 2005), and InsB18-27 (Hassainya et al., 2005; Mallone et al., 2007; Toma et al., 2005) presented in the context of HLA-A2. Several insulin B chain epitopes restricted by other HLA class I molecules (A1, A3, B8, and B18) have also been reported (Toma et al., 2005). In addition, epitopes located in the insulin A chain, as well as within the C-peptide, are found to be immunogenic in HLA-A2-transgenic mice and are targets for CD8+ T cells in diabetic patients (Hassainya et al., 2005; Jarchum et al.,

2007). Lastly, Baker et al. assayed for Granzyme B responses elicited by a complete panel of proinsulin peptides and observed a broad range of responses in recently diagnosed patients compared to healthy controls; some of these responses may arise from epitopes that have not been previously identified (Baker et al., 2008).

Not surprisingly, these studies of PBMCs in human patients have yielded a wealth of information; understanding how this information can be translated to evaluate diabetes risk or disease status is a work in progress. A potential concern arising from studying PBMCs is the unknown relevance of these cells to the repertoire of autoimmune responses that are localized in the islets or pancreatic lymph nodes (PLNs). Two recent reports aimed to address this question. Wong et al. showed that the TCRβ chain usage within the $IGRP_{206-214}$-reactive CD8+ T cell pool is similar regardless of where the cells come from: peripheral blood, pancreatic islets, or draining lymph nodes (Wong et al., 2007). Another group compared several epitopes from GAD65, proinsulin, IA-2, and IGRP that are conserved between mice and man, in terms of their recognition by CD8+ T cells isolated from the blood, spleen, lymph nodes, and pancreatic islets of individual diabetic HLA-A2-transgenic NOD mice. In this study, a significant correlation was reported between the patterns of epitope immunodominance in the different organs tested. Moreover, some of the immunodominant epitopes in humans were also found to be dominant in mice (Enee et al., 2008), demonstrating the relevance of assaying PBMCs. In addition to analyzing PBMCs, which are by far the type of samples that are most amenable to analysis in humans, epitope identification may lend itself to other less invasive ways of examining the diabetogenic responses *in vivo*. This has been demonstrated in mice, where an antigen-specific detection method (pMHC multimer staining) is combined with a magnetic resonance imaging approach to visualize antigen-specific islet inflammation in real time (Medarova et al., 2008; Moore et al., 2004).

Some of the studies discussed above incorporated data from patients at different time-points following diagnosis (recent-onset vs. long-standing or preclinical, etc.). The time parameter may be of importance, as β cell-reactive CD8+ T cells appear to shift in frequency and immunodominance as diabetes progresses (Martinuzzi et al., 2008; Trudeau et al., 2003). This dynamic quality of CD8+ T cell responses, coupled to the inherent heterogeneity in the rates at which β cell destruction occurs, results in a high degree of variability in the responses that are detected. Using this wealth of information for risk assessment purposes in humans may require the enumeration of responses to a specific/sensitive combination of a number of epitopes, or, alternatively, the enumeration of responses to epitopes that are either recognized constantly or at a specific stage of disease. Such information may also be valuable for designing treatments, as well as for defining the optimal therapeutic window.

5. THE RELATIVE CONTRIBUTION OF THE β-CELL-SPECIFIC CD8+ RESPONSE TO T1D

In terms of disease initiation, the issue of whether there actually is an inciting antigen in T1D has been a subject of much contention. It is presently unclear whether the initiation of diabetes requires autoreactive T cell responses that target multiple β cell antigens at once or primarily T cells that target a single β cell antigen, leading to epitope spreading and subsequent activation of other T cell clonotypes. To this end, insulin is believed by many to be a "triggering" antigen in mouse T1D based on observations made in proinsulin 1/2-deficient NOD mice (Moriyama et al., 2003) and in NOD mice over-expressing a pre-proinsulin 2 transgene under the control of a MHC class II promoter (I-Eα^k) in APCs (French et al., 1997). In this study, NOD mice rendered immune tolerant to proinsulin 2 were completely protected from diabetes, suggesting that autoimmunity to proinsulin was required for diabetes development (French et al., 1997). However, Jaeckel et al. followed the same approach and made observations that deviated from those of French et al.'s. In their hands, complete tolerance to proinsulin, although significantly reduced the incidence of diabetes in NOD mice, did not confer complete protection (Jaeckel et al., 2004). More recently, Krishnamurthy et al. compared diabetogenic CD8+ T cell responses and disease incidence in NOD mice over-expressing mouse pre-proinsulin 2 (NOD.PI) or IGRP (NOD.IGRP) in APCs. They observed that, unlike NOD.PI mice, NOD.IGRP mice developed insulitis and diabetes normally, compared to wild-type NOD controls (Krishnamurthy et al., 2006). Furthermore, tolerance to proinsulin significantly reduced peripheral IGRP$_{206-214}$-specific responses. Along the same line, constitutive over-expression of proinsulin in APCs of 8.3-TCR-transgenic NOD mice significantly reduced insulitis and diabetes occurrence (Krishnamurthy et al., 2008). Together, these studies may suggest that proinsulin-specific autoimmunity precedes and may contribute to unleashing IGRP-specific autoimmunity. An alternative interpretation of these data, however, is that the over-expression of proinsulin versus IGRP has affected the cross-presentation capacity or activation status of the APCs to different extents, thus resulting in the observed phenotypes. This is not a far-fetched speculation, as APCs are not the native cell types where proinsulin or IGRP are naturally expressed. Proinsulin and IGRP are two antigens that are very different in molecular nature and intracellular distribution, and are likely to induce qualitatively different responses of the cell types in which they are over-expressed. For example, over-expression of a human IGRP transgene in mouse pancreatic beta cells causes ER stress and hypo-insulinemia in the absence of islet inflammation (Shameli et al., 2007).

Unlike insulin 1-deficient NOD mice, NOD mice lacking the expression of other β cell-autoantigens, such as IA-2, IA-2β, and GAD65, develop diabetes normally (Kubosaki et al., 2004a,b; Yamamoto et al., 2004). These findings are in line with the view that multiple antigenic specificities work in concert to mediate β cell autoimmunity, a process that cannot be hindered by the selective removal of any single or combination of antigen-specific response(s). However, the contribution of individual specificities may vary. Studies in the NOD mouse showed that not all β cell-reactive CD8+ T cell responses possess the same pathogenic potential. Burton et al. compared the diabetogenicity of autoreactive T cell clones side by side using a retrogenic system and showed that β cell-reactive clonotypes isolated from NOD mice are not necessarily insulitogenic or diabetogenic on their own. In particular, the GAD- and IA-2-reactive T cell clones failed to transfer diabetes into NOD.scid hosts, despite their ability to proliferate in response to antigenic stimulation in vitro (Burton et al., 2008). Indeed, in the absence of other cell types, the pathogenicity of β cell-reactive CD8+ T cells may depend on additional factors such as the T cells' ability to home to the islets and/or to sustain local inflammation. In a study by Ejrnaes et al., a $GAD65_{501-516}$-reactive CD8+ T cell clone isolated from the spleen of a peptide-immunized NOD mouse was shown to be as capable of lysing β cells in vitro and producing inflammatory cytokines as an $mInsB_{15-23}$-specific CD8+ T cell clone. Yet, relative to the GAD-specific clone, the $mInsB_{15-23}$-specific clone was significantly more efficient at inducing diabetes when injected into neonatal NOD or NOD. scid hosts. As will be discussed later, the discrepancy between these two clones likely reflects their differential ability to recruit other inflammatory cell types (Ejrnaes et al., 2005). Of note, the diabetogenicity of any single CD8+ T cell clone is not necessarily a T cell-intrinsic property; the expression pattern and accessibility of the cognate antigens likely play an important role.

6. DEVELOPMENT AND ACTIVATION OF DIABETOGENIC CD8+ T CELLS

Studying the biology of diabetogenic T cells in wild-type NOD mice is a challenging undertaking because of their low peripheral frequency and repertoire heterogeneity. Transgenic NOD mice expressing TCRs specific for naturally occurring β cell autoantigens, hence displaying a peripheral T cell repertoire skewed towards the antigenic specificity of the transgenic TCR, have provided one solution to this limitation. As such, a number of TCR-transgenic models have been generated to study different aspects of the biology of autoreactive CD8+ T cells in detail, including their development, activation, recruitment, effector function, and regulation among others (Graser et al., 2000; Kanagawa et al., 2000; Katz et al., 1993; Schmidt et al., 1997; Tarbell et al., 2002; Verdaguer et al., 1996, 1997).

6.1. MHC class I-restricted TCR-transgenic models of T1D

To investigate the developmental biology of diabetogenic CD8+ T cells, we generated transgenic NOD mice (Verdaguer et al., 1997) expressing the TCRα and TCRβ rearrangements of a pancreatic islet-derived CD8+ T cell clone that used the conserved Vα17-Jα42 rearrangement characteristic of NRP/IGRP$_{206-214}$-reactive T cells (8.3-NOD mice) (Han et al., 2005a; Verdaguer et al., 1996). As expected, given the significant increase in the peripheral frequency of IGRP$_{206-214}$-reactive CD8+ T cells caused by expression of the transgenic TCR, 8.3-NOD mice developed an accelerated form of diabetes. Introduction of a *rag2* deficiency into this strain, to inhibit the rearrangement of endogenous TCR sequences, resulted in mice expressing a peripheral T cell repertoire exclusively composed of IGRP$_{206-214}$-reactive CD8+ T cells, devoid of CD4+ T cells. Because 8.3-NOD.*rag2*$^{-/-}$ mice cannot provide adequate CD4+ T cell help to their transgenic CD8+ T cells, they developed a lower-frequency and significantly delayed form of diabetes as compared to their *rag2*+ counterparts (Verdaguer et al., 1997).

More recently, we generated NOD mice expressing IGRP$_{206-214}$-reactive TCRs recognizing pMHC with different affinities, to investigate the impact of TCR avidity on the developmental biology and pathogenic potential of autoreactive T cells (Amrani et al., 2000a; Han et al., 2005b). We named these strains after the Vα17 element that was encoded on the transgenic TCRα chain (Vα17.4—the element used by the original 8.3-TCR–, Vα17.5 and Vα17.6); all three strains of mice expressed the same (8.3) TCRβ chain. 17.5- and 17.6-TCR+ CD8+ T cells bind IGRP$_{206-214}$/K^d tetramers with higher and lower avidity, respectively, as compared to 17.4-TCR+ CD8+ T cells. Unlike the 17.4-TCR- and 17.5-TCR-transgenic stocks, which developed an accelerated form of diabetes, the 17.6-TCR-trangenic stock was resistant to diabetes (Han et al., 2005b; Verdaguer et al., 1997). All three transgenic TCRs foster the development of IGRP$_{206-214}$-specific CD8+ T cells in the thymus, but with some notable differences. For example, unlike 17.4 and 17.6-TCR+ CD8+ T cells, 17.5-TCR+ CD8+ T cells undergo partial central and peripheral deletion, but the few cells that escape tolerance accumulate in inflamed pancreatic islets, promoting disease progression. These observations suggested that whereas central and peripheral tolerance work to limit the size of high-avidity, autoreactive T cell pools, the local inflammatory microenvironment within pancreatic islets tends to shelter clonotypes from peripheral tolerance and promote their local expansion, fostering T cell avidity maturation (Han et al., 2005b). Importantly, these observations hinted at the possibility that, for a given antigenic specificity, autoreactive CD8+ T cells might not necessarily support but rather inhibit autoimmunity, depending on their avidity for pMHC.

In another model, the TCRαβ sequences of a CD8+ T cell clone isolated from islet infiltrates of young NOD mice (DiLorenzo et al., 1998) were used to generate NOD.AI4αβ TCR-transgenic animals (Graser et al., 2000). The AI4 TCR recognizes residues 138–146 of a widely expressed antigen, Dystrophia Myotonica Kinase ($DMK_{138-146}$), in the context of H-2D^b (Lieberman et al., 2004). These mice develop an accelerated form of diabetes that is CD4+ T cell help-independent (Graser et al., 2000). Expression of TCRαβ transgenes derived from another autoreactive CD8+ T cell clone of unknown antigenic specificity also fostered disease acceleration, although in this case, and similar to what we saw in 17.5-TCR-transgenic NOD mice (Han et al., 2005b), there was significant deletion of transgenic thymocytes (Kanagawa et al., 2000).

Another approach to studying β cell autoimmunity involves the expression of neo-self antigens in pancreatic β cells under the control of the rat insulin promoter (RIP). In these models, diabetes is induced by active immunization (i.e., viral infection), to elicit an immune response targeting the transgenic neo-self antigen. In mice expressing lymphocytic choriomeningitis virus glycoprotein (LCMV-GP) or nucleoprotein (LCMV-NP) transgenes in β cells (Oldstone et al., 1991), viral infection is used to overcome the "ignorance" of the transgenic neo-antigen by peripheral autoreactive T cells, and thus triggers β cell autoimmunity (Ohashi et al., 1991). In contrast to the RIP-LCMV-GP model, double transgenic mice expressing the influenza virus hemagglutinin in pancreatic β cells (RIP-HA) and a $HA_{512-520}/K^d$ specific TCR transgene (CL-4) developed an aggressive form of juvenile diabetes that was preceded by CD8+ T cell infiltration in the pancreatic islets (Morgan et al., 1996). Similarly, rapid onset of diabetes was observed when a RIP-ovalbumin (RIP-OVA) transgene was co-expressed with a transgenic OVA-specific, K^b-restricted TCR (OT-1) (Blanas et al., 1996). Differences in genetic backgrounds, pMHC-binding affinities of the transgenic TCRs, and levels, timing and anatomical geography of transgene expression, may very well account for differences in outcome.

6.2. Activation of diabetogenic CD8+ T cells

The activation of autoreactive CD8+ T cells requires the presentation of autoantigenic pMHC class I complexes along with costimulatory molecules on the surface of professional APCs, such as DCs. Target cells do not express costimulatory molecules and are thus unlikely to elicit an autoimmune response. Cross-presentation is a major pathway through which naive autoreactive CD8+ T cells can access autoantigenic pMHC complexes in the draining lymph nodes (Heath and Carbone, 2001); DCs that acquire cell-associated antigens (such as those released during apoptotic cell death) can present them through the MHC class-I pathway (Albert et al., 1998). In the NOD mouse, some evidence suggest that the initial priming of CD8+ T cells occurs in the PLNs, as surgical removal of PLNs at three weeks of age almost completely protected the mice from the development of insulitis and diabetes (Gagnerault et al., 2002). Similarly, both islet-specific CD4+ (Hoglund et al., 1999; Turley et al., 2003) and CD8+ T cells (Zhang et al., 2002) proliferate specifically in the PLNs upon adoptive transfer into NOD mice, indicating that β cell antigens are presented primarily by APCs residing in the PLNs.

How β cell antigens come to be presented by APCs in the PLNs is still unclear. We have previously shown that abrogation of MHC class-I expression on pancreatic β cell delayed the recruitment and reduced the retention of $IGRP_{206-214}$-reactive CD8+ T cells in the islets, but did not blunt the proliferation and activation of these cells in the PLNs, suggesting that cross-presentation of β cell antigens in the PLNs can proceed in a T cell-independent manner, i.e., without a preceding CD8+ T cell attack on β cells (Yamanouchi et al., 2003). Initial β cell autoantigen shedding can be ascribed to the neonatal remodeling of pancreatic islets resulting from a wave of physiologic β cell apoptosis peaking at two weeks of age in mice (Trudeau et al., 2000). This wave of β cell apoptosis may enhance the capture of autoantigens (possibly segregated within apoptotic bodies) by professional APCs and their presentation in the PLNs. In fact, induction of β cell apoptosis either with low doses of the β cell toxin streptozotocin (Turley et al., 2003; Zhang et al., 2002) or through injection of diabetogenic CTLs (Kurts et al., 1998) enhanced the priming of diabetogenic T cells in the PLNs. Similarly, intra-pancreatic injection of dead β cells induced proliferation of autoreactive T cells in the PLNs at an age at which these T cells would not spontaneously proliferate (Turley et al., 2003).

6.3. The requirement for CD4+ T cell help

Diabetogenic CD8+ T cells may require CD4+ T cell-help to reach their maximum diabetogenic potential. For instance, in the absence of CD4+ T cells, naïve $IGRP_{206-214}$-reactive CD8+ T cells in rag2-deficient 8.3-TCR-transgenic NOD mice cannot efficiently home to the islets, resulting in a retardation in diabetes onset as compared to rag2-competent 8.3-TCR-transgenic NOD mice (Verdaguer et al., 1997). In the RIP-mOVA model of autoimmune diabetes, deletional tolerance limits the ability of small numbers of OVA-specific CD8+ T cells (OT-I) to cause diabetes. On the other hand, co-transfer of OVA-specific CD4+ helper T cells with OT-I T cells impaired peripheral tolerance of OT-I cells and favored diabetes induction (Kurts et al., 1997). The requirement for CD4+ T cell-help in these models is consistent with another study, where CD8+ splenocytes from diabetic NOD donors could not transfer diabetes into NOD.scid mice in the absence of co-transferred CD4+ T cells (Christianson et al., 1993).

In contrast to the above findings, diabetogenic CD8+ T cells in the AI4-TCR-transgenic model continued to cause diabetes at an accelerated rate in the absence of CD4+ T cell-help (Graser et al., 2000), suggesting that CD4+ T cell-help is not an absolute requirement for all diabetogenic CD8+ T cells.

Helper CD4+ T cells fuel the activation of CD8+ T cells by activating APCs, most probably through ligation of CD40 on APCs by CD154 on CD4+ T cells (Bennett et al., 1998). CD40 engagement induces the upregulation of costimulatory molecules such as B7 and ICAM-1 on APCs and increases the production of proinflammatory cytokines such as TNF-α and IL-12 (Grewal and Flavell, 1998). In the absence of CD40/CD154 interactions, NOD mice are completely protected from insulitis and diabetes (Amrani et al., 2002; Green et al., 2000). We have previously demonstrated that expression of CD154 is an absolute requirement for the activation of a population of diabetogenic CD4+ T cells, and that such defect cannot be overcome by strategies that bypass CD154-mediated activation of APCs such as CpG-DNA, agonistic anti-CD40 mAb, NOD CD4+ T cells, or expression of B7.1 on β cells. In the case of diabetogenic CD8+ T cells, the CD154-deficiency did not abrogate the diabetogenic potential of CD8+ T cells in $rag2^{-/-}$ 8.3-TCR-transgenic NOD mice, and activation of APCs by CpG DNA, agonistic anti-CD40 mAb, or NOD CD4+ T cells increased the diabetogenic potential of these cells (Amrani et al., 2002). Interestingly, the CD154-deficiency completely prevented diabetes in $rag2$-competent 8.3-TCR transgenic NOD mice, which contained endogenous CD154-deficient CD4+ T cells. We showed that protection from diabetes in this model was mediated by CD4+ CD25+ regulatory T cells, whose suppressive function on DCs prevailed in the absence of an activatory signal mediated by CD40-CD154 interactions from CD4+ CD25- helper T cells. This inhibitory effect by CD4+ CD25+ T cells required CTL-associated antigen 4 (CTLA-4), and was reversed by DC activation through injection of CpG-DNA or agonistic anti-CD40 mAb (Serra et al., 2003). In another study, transgenic expression of TNF-α on islets promoted the cross-presentation of islet antigens to CD8+ T cells in a CD40-CD154-independent manner, bypassing the requirement for CD4+ T cell-help in APC maturation (Green et al., 2000).

6.4. Inflammatory signals augment diabetogenic CD8+ T cell activation

In addition to the differential requirements for CD4+ T cell-help, there are multiple factors that determine the fate of cross-presentation and activation of CD8+ T cells. These include the level of antigen expression, the nature of antigen (soluble versus cell-associated), type of tissue injury (apoptotic versus necrotic cell death), the nature of APCs (B cells, macrophages or DCs) and their maturation status, the presence of "danger or stranger signal", the frequency of the effector cells, and the presence of immune-complex forming antibodies (Heath and Carbone, 2001).

The fate of the immune response against a specific antigen depends on the milieu in which the antigen is presented. In the presence of proinflammatory cytokines and danger signals, antigen presentation by APCs will lead to immunity, while in a non-inflammatory environment, T cell tolerance normally ensues. In RIP-HA-transgenic mice, presentation of neo-self antigen to adoptively transferred CL-4 CD8+ T cells in a non-inflammatory condition resulted in the deletion of CL-4 CD8+ T cells rather than in the induction of insulitis or diabetes. The rate of tolerance was inversely correlated with the frequency of CD8+ T cells and directly proportional to the amount of antigen presented in draining lymph nodes. That is, homozygous RIP-HA$^{+/+}$ mice were rendered tolerant more efficiently compared to RIP-HA-heterozygous mice. However, RIP-HA-mediated tolerance could be offset by immunization with influenza virus, which acted as a source of inflammatory signals, bringing about the rapid onset of diabetes (Morgan et al., 1999).

Target cell death enhances cross-presentation of tissue-specific antigen through the release of so-called endogenous adjuvants, such as heat shock proteins (HSPs) and other Toll-like receptor (TLR) ligands, as well as uric acid. These molecules promote DC activation and shift the immune response from tolerance towards immunity (Rock and Shen, 2005). A role for danger signals coming from endogenous adjuvants released during necrotic cell death has been implicated in the initiation and progression of systemic autoimmune disorders such as systemic lupus erythematosus (Marshak-Rothstein, 2006). In T1D, these have been investigated in a number of studies. HSP70 promoted diabetes in the RIP-LCMV-GP/P14 TCR-transgenic (a LCMV-GP-specific TCR) model. While injection of HSP70 or GP33 (the cognate peptide for the P14-TCR) alone did not induce diabetes, co-injection of HSP70 with GP33 enhanced diabetes development in these mice. HSP70 enhanced the ability of DCs to activate CD8+ T cells and converted them from tolerogenic into immunogenic DCs (Millar et al., 2003). In another study using the same model, TLR ligands were shown to promote the progression of insulitis to diabetes through IFN-α-mediated upregulation of MHC class-I on pancreatic islet cells (Lang et al., 2005).

Necrotic and apoptotic cells are known to release other proinflammatory molecules such as uric acid. Elimination of uric acid from the extracellular space significantly reduced the proliferation of OT-1 CD8+ T cells (specific for a peptide in ovalbumin, OVA) in the PLNs of RIP-mOVA transgenic mice. This effect could be reversed by injecting an agonistic anti-CD40 mAb, suggesting that CD4+ T cell-help through CD40-CD40L interactions might compensate for the lack of danger signals (Shi, Galusha and Rock, 2006).

6.5. Defective immune tolerance fuels the activation of diabetogenic CD8+ T cells

The NOD mouse is considered a model of immune-dysregulation due to its manifold abnormalities in both central and peripheral tolerance pathways (Anderson and Bluestone, 2005). These stem from both T cell-intrinsic defects, as well as the defects that contribute to an overall immuno-activatory environment. In the thymus, T cell-intrinsic defects impair the negative selection of autoreactive thymocytes (Kishimoto and Sprent, 2001). For instance, defective induction of proapoptotic gene expression (Liston et al., 2004, 2007), or non-MHC gene-controlled down-regulation of TCRs in a T1D-prone genetic background (i.e., NOD) (Choisy-Rossi et al., 2004; Serreze et al., 2008) may promote the escape of diabetogenic T cells into the periphery.

In the periphery, a number of tolerance defects are also known to contribute to aberrant CD8+ T cell responses. For instance, macrophages in NOD mice are defective in clearing apoptotic cells (O'Brien et al., 2002, 2006), a phenomenon that is also observed in a number of other autoimmune settings (Hanayama et al., 2004; Viorritto et al., 2007). Repeated induction of apoptosis led to production of antinuclear autoantibodies (ANA) (a phenomenon associated with systemic autoimmunity) in NOD mice but not in non-autoimmune prone strains (O'Brien et al., 2006). Therefore, defective clearance of apoptotic cells may augment cross-presentation to autoreactive T cells through enhancing the access of autoantigens by DCs.

Additional contributions to NOD immune dysregulation may come from functional or numerical defects of immunoregulatory cell types, an example of which is the classical CD4+ CD25+ Treg subset (Gregori et al., 2003). In this respect, we have shown that a CD8+ (and CD4+) T cell-intrinsic defect in production of the immunoregulatory cytokine Interleukin-2 (IL-2) is in part responsible for the impaired regulatory T cell function seen in NOD mice (Yamanouchi et al., 2007). Invariant NKT (iNKT) cells also fall into this cell category, as demonstrated in a number of studies (Novak et al., 2007). Serreze and colleagues showed that transfer of diabetes into sublethally irradiated NOD mice with splenocytes from NOD.$rag1^{null}$-AI4 TCR transgenic mice could be prevented by pretreatment of recipient NOD mice with α-galactosylseramide (α-GalCer, a synthetic ligand for iNKT cells). A proposed mechanism for α-GalCer-mediated protection is that iNKT cells activated by α-GalCer enhance the recruitment of tolerogenic DCs to the PLNs, inducing deletion and/or anergy of diabetogenic CD8+ T cells (Chen et al., 2005). Here, it should be mentioned that observations that are at odds with these findings have also been noted. In the RIP-HA/CL4-TCR-transgenic model of diabetes, activating or expanding iNKT cells enhanced the activation and differentiation of CL4-TCR CD8+ T cells into effector CTLs, resulting in an increase in diabetes incidence (Griseri et al., 2005). These differences in the functional role of iNKT cells on CD8+ T cell-mediated diabetogenesis might originate from the differences in the genetic backgrounds of the diabetes models used and/or differences in the affinities of the TCRs that were studied.

6.6. Cross-tolerance of diabetogenic CD8+ T cells

Following cross-presentation, an alternative outcome, termed "cross-tolerance", can ensue, where self-antigens cross-presented to cognate CD8+ T cells by APCs, particularly DCs, induce T cell tolerance instead of activation (Heath and Carbone, 2001; Luckashenak et al., 2008). Negative co-stimulatory molecules of the CD28 family, such as CTLA-4 and programmed death 1 (PD-1), have been shown to be critical for peripheral cross-tolerance (Keir et al., 2008; Probst et al., 2005). CTLA-4 is an inhibitory receptor expressed on T cells after TCR-ligation, and engages the co-stimulatory molecules CD80 and CD86 on DCs (Perez et al., 1997). It is also constitutively expressed on CD4+ CD25+ Tregs and is critical for their functional activation. Loss of CTLA-4 results in hyper-lymphoproliferation and multi-organ autoimmunity (Tivol et al., 1995; Waterhouse et al., 1995), and in vivo blockade of CTLA-4 results in chronic autoimmunity (Perrin et al., 1996; Probst et al., 2005). Not surprisingly, CTLA-4 is genetically associated with murine and human T1D (Howson et al., 2007; Ueda et al., 2003).

PD-1 is another such inhibitory receptor. Expressed on activated T cells, B cells, and myeloid cells, PD-1 is implicated in both central and peripheral tolerance (Keir et al., 2008). The interaction of PD-1 with its ligands, PD-L1 (B7-H1) and PD-L2 (B7-DC) on multiple immune system cell types, including DCs, macrophages, and lymphocytes, as well as parenchymal cells such as the pancreatic β cells, delivers an abortive signal that dampens T cell activation. Abrogating the PD-1/PD-L1 interaction, either by genetic manipulation or antibody blockade, leads to enhanced CD8+ T cell-mediated autoimmunity (Keir et al., 2006, 2007; Latchman et al., 2004; Nishimura et al., 1999; Salama et al., 2003). In the context of T1D, studies have shown that the interaction of PD-1 on CD8+ T cells and PD-L1 on both DCs and pancreatic β cells is important for inducing and maintaining peripheral CD8+ T cell tolerance to β cell antigens (Keir et al., 2007; Wang et al., 2005). For instance, PD-L1 blockade or PD-1 deletion leads to accelerated diabetes in NOD mice (Ansari et al., 2003). Keir and coworkers used the adoptive transfer model of OVA-specific OT-1 CD8+ T cells in RIP-OVA recipient mice to study the effects of the PD-1/PD-L1 interaction on cross-priming and tolerance and found that the involvement of the PD-1/PD-L1 pathway in cross-tolerance is two-fold. In the draining lymph nodes, the interaction of PD-1 with PD-Ls on DCs negatively regulates OT-1 T cell activation. PD-1-deficient OT-1 CD8+ T cells have a lower threshold of activation and can efficiently transfer diabetes to RIP-OVA hosts, unlike their PD-1-competent counterparts. In the target tissue (i.e., pancreas), PD-L1 deficiency exacerbates OT-1 T cell-mediated diabetes, suggesting a role for PD-L1 in protecting the target organ against reactivation of autoreactive CD8+ T cells. Another series of adoptive transfer experiments using β cell antigen-specific, diabetogenic CD4+ and CD8+ T cells found that PD-L1 negatively regulates the priming of diabetogenic T cells in the PLN in early stages of disease, while PD-L1 in islets diminishes islet destruction at a later phase in diabetes progression (Guleria et al., 2007). Of note, similar to CTLA-4, the PD1/PD-L1 pathway has been implicated in therapeutic applications for autoimmune diseases. Fife and coworkers demonstrated that systemic administration of FcR-non-binding anti-CD3 monoclonal antibodies or antigen-coupled APCs reversed recent onset diabetes in NOD mice and maintained long-term protection thereafter, in a PD-1 and PD-L1-dependent manner (Fife et al., 2006). As another example, stimulation of PD-1 with a dimeric PD-L1 Ig fusion protein, in conjunction with CD154 co-stimulation blockade, prolonged the survival of islet allografts (Gao et al., 2003).

7. RECRUITMENT OF DIABETOGENIC CD8+ T CELLS TO ISLETS

The inflammatory milieu of the islets can impact the recruitment of autoreactive CD8+ T cells through the expression of chemokines and homing ligands and determine the fate of immune responses against β cells. The homing of autoreactive cells to the islets is impaired in the absence of IFN-γ, as it is required for T cell diapedesis and penetration into islets (Savinov et al., 2001). In the LCMV-induced model of autoimmune diabetes in RIP-LCMV-GP transgenic mice, it has been demonstrated that, in response to islet inflammation, β cells produce the chemokines IFN-γ inducible protein-10 (IP-10 or CXCL10) and CXCL9, which in turn enhance the recruitment of autoreactive CD8+ T cells to the site via chemokine receptor CXCR3. In fact, CXCR3−/− mice displayed delayed onset of diabetes in this model (Frigerio et al., 2002). Another study using the same model showed an early and substantial expression of IP-10 but not CXCL9 or CXCL11 in pancreatic islets. The administration of blocking antibodies against IP-10 (but not the administration of antibodies against other chemokines) abrogated type-1 diabetes in this model (Christen et al., 2003). Furthermore, CTL clones specific for two different β cell autoantigens with the same cytolytic, cytokine, and homing profiles exhibited differential diabetogenicity when transferred into NOD mice. The more diabetogenic clone expressed higher levels of IP-10, which promoted the recruitment of T cells and macrophages to the islets (Ejrnaes et al., 2005).

Activation of autoreactive T cells is associated with changes in the expression of surface markers that mediate pancreatic islet homing.

In general, T cell activation induces the down-regulation of L-selectin (CD62L) and increases the expression of adhesion molecules such as CD44 and the integrin very late activation antigen-4 (VLA-4), facilitating their homing to the inflamed target organs. Moreover, interactions with DCs presenting tissue-specific antigens in the draining LNs can induce the expression of specific chemokine receptors and adhesion molecules on T cells that are required for their preferential homing to the target organ (Dudda and Martin, 2004). The integrins α4β1 (VLA-4) and α4β7 (LPAM-1) have been implicated in the homing of T cells to pancreatic islets and in diabetogenesis. In the RIP-mOVA/OT-1 model of T1D, antigen presentation in the PLNs led to expression of the integrin α4β1 on autoreactive T cells and imprinted upon them a preference for islet homing. Furthermore, blockade of VCAM-1 (the ligand for the integrin α4β1) resulted in diabetes protection in this model (Hanninen et al., 2007). Another study showed that CTLs recognizing Insulin B_{15-23} could home to pancreatic islets and cause diabetes without the need for preexisting islet inflammation. The authors provided evidence that endothelial cells in the islet vasculature take up secreted insulin and cross-present the antigenic epitope to CTLs in the context of K^d. The secondary lymphoid organ chemokine SLC (CCL21), was required for such adhesion and homing through its interaction with the chemokine receptor CCR7 on T cells (Savinov et al., 2003b).

8. MECHANISMS OF β-CELL CYTOTOXICITY IN T1D

Cytotoxic T cells kill target β cells through direct or indirect mechanisms. Direct killing occurs through either the Fas or granule exocytosis/perforin pathways and requires that T cells recognize autoantigens presented by MHC molecules on the surface of β cells. Indirect killing involves soluble factors such as proinflammatory mediators produced by T cells, APCs, or even β cells.

8.1. Fas versus granule exocytosis

The relative role of Fas or granule exocytosis/perforin pathways in the direct destruction of β cells remains controversial despite numerous studies conducted to date. Perforin-deficient NOD mice develop insulitis but present with significantly reduced incidence of diabetes, suggesting that perforin is required for β cell destruction (Kagi et al., 1997). Similar results were obtained using the LCMV-induced model of T1D (Kagi et al., 1996). Using another model where diabetes was induced by adoptive transfer of influenza HA-specific CD8+ T cells into RIP-HA mice, disruption of perforin-mediated killing delayed the onset of diabetes, and induction of diabetes required 30-times more CD8+ T cells. On the other hand, disruption of Fas/Fas-ligand (FasL) pathway had little impact on diabetes incidence in this model (Kreuwel et al., 1999). In contrast to these findings, we showed that in the 8.3–TCR-transgenic NOD model, introduction of a perforin deficiency did not attenuate diabetes and that $IGRP_{206-214}$-reactive CD8+ T cells killed β cells primarily via Fas (Amrani et al., 1999). A more recent report, however, has suggested that both Fas and perforin might be involved in this model (Dudek et al., 2006).

The role of Fas in β cell killing has been investigated in many other studies. Fas-deficient NOD.lpr mice developed neither insulitis nor diabetes; and adoptive transfer of splenocytes from diabetic NOD mice induced diabetes in normal NOD mice but not NOD.lpr mice (Itoh et al., 1997). However, the presence of multiple immunological abnormalities in NOD.lpr mice raised the concern that these observations might be the consequence of a T cell defect in these animals. In fact, it was shown that T cells in the NOD.lpr mice express high levels of FasL, causing the rapid destruction of any adoptively transferred T cells via Fas-FasL interactions (Kim et al., 2000).

More recently, studies conducted in mice lacking Fas exclusively in β cells provided a better understanding of this pathway in the context of an otherwise unaltered immune system. β cell-specific over-expression of a dominant negative Fas (Savinov et al., 2003a) or a dominant negative Fas-Associated Death Domain (FADD, involved in apoptosis signaling from death receptors) (Allison et al., 2005) delayed spontaneous diabetes in NOD mice. However, Fas deficiency did not protect β cells from autoimmune destruction, as Fas-deficient islets were readily rejected when grafted into diabetic NOD mice (Allison and Strasser, 1998), or upon adoptive transfer of splenocytes from diabetic mice (Pakala et al., 1999).

Taken together, these studies suggest that both Fas and perforin pathways contribute to CD8+ T cell-mediated destruction of pancreatic β cells. The relative contribution of these pathways may vary, depending on the avidity of CTL clones for their respective antigens. Some evidence suggests that Fas-mediated cytotoxicity may be elicited at a lower threshold of avidity than perforin-mediated cytotoxicity (Kessler et al., 1998). This is an attractive model, as it provides an explanation for the observations that perforin-deficientNOD mice develop insulitis and not diabetes (Amrani et al., 1999; Kagi et al., 1997), and that Fas-deficient islet grafts are destroyed in diabetic hosts (Allison and Strasser, 1998) or after transfer of splenocytes from diabetic mice (Pakala et al., 1999). In fact, we have evidence that avidity maturation of $IGRP_{206-214}$-reactive CD8+ T cells during progression of disease in NOD mice is associated with a shift from Fas to perforin-mediated killing (Qin et al., 2004). Ultimate resolution of this controversy will require studying the dynamics of beta cell killing in mice in which Fas expression can be exclusively removed in β cells and in a time-controlled manner.

8.2. TNF-α, TNFR, IFN-γ, and IL-1

In addition to perforin and Fas, a third mechanism of cell-cell contact-dependent β cell killing involving membrane-bound TNF was reported in double transgenic mice expressing CD80 and TNF on pancreatic β cells (RIP-TNF/CD80 mice). These mice developed a form of CD8+ T cell-mediated diabetes that was independent of Fas and perforin, but required the expression of TNF receptor II (TNF-RII). It was suggested that membrane TNF/TNF-RII interactions might be involved in the killing process, although these data are hard to interpret due to defective immune system development consequent to the TNF mutation (Herrera *et al.*, 2000).

Soluble factors such as proinflammatory cytokines, free radicals and reactive oxygen intermediates have been implicated in the indirect killing of β cells in T1D. This is consistent with the fact that diabetogenic CD4+ T cells can induce diabetes in the absence of CD8+ T cells despite the fact that pancreatic β cells do not express MHC class II molecules. Activated CD8+ T cells produce large amounts of IFN-γ and TNF-α, which are directly or indirectly involved in β cell destruction. Systemic administration of TNF-α enhanced diabetes when given to young NOD mice (Yang *et al.*, 1994). Similarly, neonatal expression of TNF-α on β cells accelerated diabetes development (Green *et al.*, 1998). β cells constitutively express low levels of TNF-RI, but can be induced to express both TNF-RI and TNF-RII during insulitis (Walter *et al.*, 2000). Moreover, lack of TNF-RI on β cells protects them from autoimmune destruction (Pakala *et al.*, 1999).

Fas expression is normally undetectable on NOD β cells but can be upregulated in response to inflammatory signals such as IFN-γ and IL-1 (Darwiche *et al.*, 2003; Thomas *et al.*, 1999). We have shown that IFN-γ, IL-1α, and IL-1β induce upregulation of Fas on islet cells and facilitate their killing by diabetogenic CD4+ T cells (Amrani *et al.*, 2000b). In addition, these cytokines enhance the expression of iNOS and NO production, resulting in β cell dysfunction and damage (Thomas *et al.*, 2002). Furthermore, IFN-γ induces the upregulation of MHC class I expression on NOD β cells (Thomas *et al.*, 1998), and in the presence of other inflammatory cytokines can induce β cell death (Seewaldt *et al.*, 2000). Collectively, these studies demonstrate that proinflammatory mediators not only predispose β cells to Fas or perforin-mediated lysis through increasing the expression of FasL and MHC class I, but may also be directly involved in the killing process.

The role of cytokines in diabetes development was further confirmed in a study using 8.3- TCR transgenic mice over-expressing the suppressor of cytokine signaling-1 (SOCS-1) specifically in β cells. This approach inhibits β cell responses to cytokines, such as IFN-γ, that signal via the JAK/STAT pathway. Despite the presence of insulitis, these mice were completely protected from diabetes, suggesting that inhibiting the

IFN-γ-mediated upregulation of Fas and MHC class I on pancreatic β cells protects them from destruction by CTLs (Chong et al., 2004).

Taken together, these studies suggest that Fas and perforin are the major pathways for CTL-mediated killing of β cells, while their function might be amplified in the presence of inflammatory cytokines and other soluble mediators. The relative contribution of these pathways may depend on the affinity/avidity of the TCR/T cell–pMHC/APC interaction, the stage of disease, and the inflammatory milieu within the islets.

9. INDUCTION OF IMMUNOLOGIC TOLERANCE IN DIABETOGENIC CD8+ T CELLS

The NOD mouse model has been exploited as a vehicle for testing multiple antigen-specific immunotherapies for T1D, which range from the administration of intact proteins and antigenic peptides to antigen-encoding DNA, and through systemic routes in the presence or absence of adjuvants, to more local ones such as the gut and nasal mucosa (Filippi et al., 2005). Autoantigens coupled to carrier APCs such as splenocytes (Fife et al., 2006) or loaded onto tolerogenic DCs (Lo et al., 2006) have also demonstrated efficacy in preventing T1D. These protocols were shown to dampen β cell autoimmunity by inducing deletional tolerance and/or anergy in autoreactive T cells, or by generating antigen-specific Tregs.

To our knowledge, relatively few studies to date have documented the targeting of CD8+ T cells to achieve β cell-specific tolerance in NOD mice (Table 4.1). In one study, deletional tolerance of a diabetogenic CD8+ T cell clone was achieved by targeting its cognate antigenic mimotope peptide to tolerogenic DCs with an antibody against the surface marker DEC-205, an endocytic receptor expressed on DCs (Mukhopadhaya et al., 2008).

We and others have shown that targeting specific populations of diabetogenic CD8+ T cells with cognate peptide ligands or altered peptide ligands (APLs) is an effective means of blunting T1D progression in animal models. Injection of a soluble LCMV-GP peptide (GP33) into LCMV-GP-specific TCR transgenic mice induced an initial expansion of LCMV-specific T cells followed by subsequent deletion and anergy (Kyburz et al., 1993). Repeated intra-peritoneal administration of the GP33 peptide induced tolerance in diabetogenic CD8+ T cells and completely suppressed LCMV-induced diabetes in RIP-LCMV-GP mice (Aichele et al., 1994). Similarly, in the RIP-HA×CL4-TCR transgenic model of T1D, intravenous administration of an immunogenic influenza virus hemagglutinin (HA) peptide into CL4-TCR transgenic mice induced activation and subsequent deletion or unresponsiveness of CL4 CD8+ T cells and delayed the onset of spontaneous diabetes (Bercovici et al., 2000).

A superagonistic APL of an HA epitope was compared to its wild-type counterpart in terms of its efficacy in blocking CTL-mediated diabetes in RIP-HA mice; it was found that the superagonistic peptide was as effective as wild-type peptide in deleting CTLs, but significantly more effective at suppressing the cytotoxicity of CTLs (Hartemann-Heurtier et al., 2004).

In a more physiological setting, peptides derived from disease-relevant antigens, Insulin 2 and GFAP have also shown efficacy in blunting diabetes when co-administered with incomplete Freund's adjuvant (IFA) into prediabetic NOD mice (Martinez et al., 2003; Tsui et al., 2008). Martinez et al. showed that peptide immunotherapy with the CTL epitope, mIns2 B25-C2 reduced diabetes incidence (Martinez et al., 2003); Tsui et al. tested two GFAP peptides and found that immunotherapy with $GFAP_{79-87}$ afforded significant diabetes protection, while immunotherapy with another peptide, $GFAP_{253-261}$, did not (Tsui et al., 2008). These discrepancies reflect the differential requirements (e.g., dose, timing, route) for the induction of tolerance against each antigenic specificity, using this particular method (Tsui et al., 2008).

We have shown that repeated systemic administration of a soluble $IGRP_{206-214}$ mimotope, NRP-A7, effectively prevented diabetes in prediabetic NOD mice (Amrani et al., 2000a). To further understand the mechanism underlying the observed protective effect, we used APLs of $IGRP_{206-214}$-reactive CD8+ T cells in a wide range of affinities and doses. High doses of a low-affinity APL (NRP-I4) and intermediate doses of an intermediate-affinity APL (NRP-A7) afforded complete protection against diabetes. Furthermore, this protection was associated with the deletion of high-avidity $IGRP_{206-214}$-reactive CD8+ T cells, together with the expansion and local accumulation of their low-avidity counterparts in the pancreatic islets. Interestingly, disease protection was abrogated at high doses of NRP-A7, as well as at all doses of NRP-V7 (a superagonistic mimic of $IGRP_{206-214}$). Analyses of the islet infiltrates of non-protected mice indicated near-complete deletion of all $IGRP_{206-214}$-reactive CD8+ T cells, together with the expansion of CD8+ T cells recognizing other subdominant IGRP epitopes (Han et al., 2005a). Thus, we proposed that, under conditions that spare low-avidity clonotypes, APL treatment promotes the occupation of the islet lymphocyte "niche" by low-avidity, non-pathogenic CD8+ clonotypes, thereby physically hindering the access of β cells by other pathogenic clonotypes (Fig. 4.1).

However, physical blockade may not be the only explanation for the observed protective effect. It is reasonable to suspect that the low-avidity, autoreactive clonotypes expanded in this manner are immunoregulatory in nature. To investigate the anti-diabetogenic potential of low-avidity, autoreactive CD8+ T cells, we studied a transgenic NOD strain expressing a low-affinity $IGRP_{206-214}$-reactive TCR (17.6-TCR). These mice foster

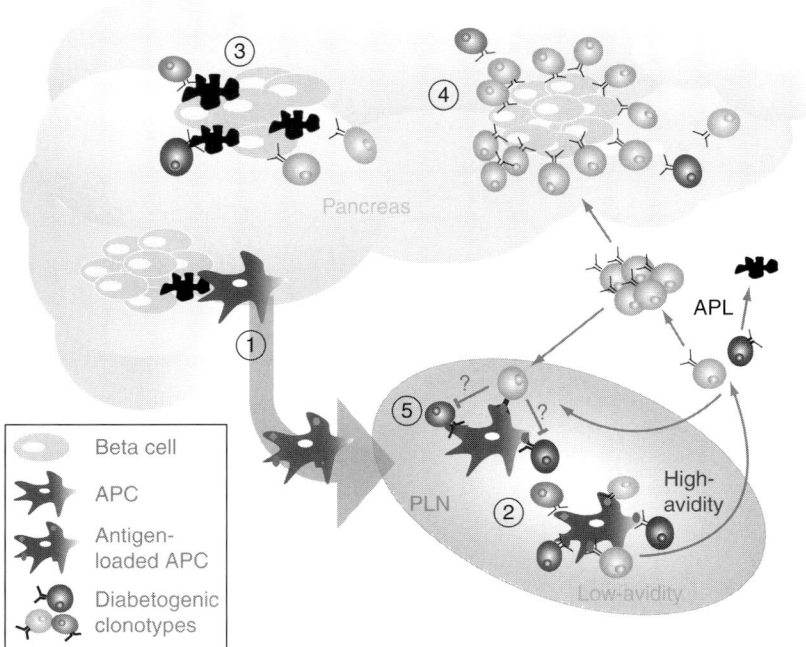

FIGURE 4.1 Protection from diabetes through selective expansion of low-avidity autoreactive T cells: Physiologic β cell death initiates autoimmunity by providing autoantigens for presentation by DCs in the PLNs (1). The inflammatory milieu in the PLN promotes the activation of autoreactive T cells specific for multiple β cell antigens (2) Autoreactive T cells home to the pancreatic islets, where they mediate β cell destruction (3). Antigen-specific immunotherapy (e.g., APLs, at optimal doses and affinities) can selectively delete high-avidity pathogenic clonotypes and expand low-avidity non-pathogenic ones that either passively protect β cells from other pathogenic clonotypes (4), or exhibit active suppression in the PLNs through a mechanism that may involve autoantigen-loaded DCs (5). (See Plate 7 in Color Plate Section.)

the development of two phenotypically distinct CD8+ T cell populations: a subset of diabetogenic $IGRP_{206-214}$-reactive CD8+ T cells that binds $IGRP_{206-214}/K^d$ tetramer with high avidity and expresses high levels of transgenic TCRs, and a subset of CD8+ T cells that binds $IGRP_{206-214}/K^d$ tetramer with low-to-intermediate avidity and expresses both transgenic and endogenous TCRs. Interestingly, despite the presence of tetramer-high, diabetogenic CD8+ T cells, these mice do not develop diabetes. Protection from diabetes is dependent on the presence of low-avidity CD8+ T cells, as the selective removal of these cells by introducing a *rag2* gene deficiency (which does not impair the development of the high-avidity, pathogenic subset) results in loss of protection from diabetes (A. Shameli, S. Tsai, and P. Santamaria, unpublished observations).

Our recent studies demonstrated that low-avidity IGRP$_{206-214}$-reactive CD8+ T cells in 17.6-TCR-transgenic NOD mice spontaneously differentiate into a population of memory CD8+ T cells, that are absent in the higher-affinity IGRP$_{206-214}$-reactive TCR transgenic (8.3) NOD mice. These low-avidity autoreactive CD8+ T cells suppress the activation and proliferation of their naïve high-avidity counterparts in a contact-dependent manner, through a process that is mediated by killing of autoantigen-loaded APCs in the secondary lymphoid organs (A.Shameli, S. Tsai, and P. Santamaria unpublished data). Excitingly, we have been able to selectively expand this population of low-avidity IGRP$_{206-214}$-reactive CD8+ T cells in TCR-non-transgenic NOD mice using a strategy that specifically targets the IGRP$_{206-214}$-reactive CD8+ T cells (our unpublished data). The end result is the induction of active tolerance, leading to long-lasting diabetes protection in prediabetic animals, as well as the restoration of normoglycemia in recent-onset diabetic animals. We are now convinced that, with the right conditions, antigen-specific induction of immune tolerance is a feasible approach for treating T1D, and possibly other CD8+ T cell-mediated autoimmune diseases.

10. CONCLUDING REMARKS

Recent findings have highlighted the importance of autoreactive CD8+ T cells in the pathogenesis of T1D and other organ-specific autoimmune disorders. In the NOD mouse, unlike the case in diabetes-resistant strains, defects in immune tolerance favor the activation of autoreactive CD8+ T cells, which in turn effect β cell cytotoxicity and contribute to islet inflammation. The unique roles that CD8+ T cells play in T1D make them particularly attractive targets for therapeutic intervention. Importantly, studies using the NOD model have yielded proof-of-concept that restoration of tolerance to β cells can be achieved through antigen-specific manipulation of autoreactive CD8+ T cell subsets. Our studies have demonstrated that low-avidity autoreactive (but anti-diabetogenic) CD8+ T cells arise spontaneously, and that selective expansion of these cells can effectively dampen β cell autoimmunity. Others have also shown that combining an antigen-specific therapy with systemic-based immune-modulation may enhance the success of therapy. Presently, concerted efforts focus on uncovering disease-relevant β cell epitopes for the purpose of designing better diagnostic tools and identifying potential targets for immunotherapy. Together, these advancements foster hope for finding a cure for T1D.

ACKNOWLEDGMENTS

We thank P. Serra, J. Wang, J. Yamanouchi, and Y. Yang for reading the manuscript. Some of the work reviewed here was supported by grants from the Canadian Institutes of Health Research, the Natural Sciences and Engineering Research Council of Canada, the Canadian Diabetes Association, and the Juvenile Diabetes Research Foundation. A.S. and S.T. are supported by studentships from the Alberta Heritage Foundation for Medical Research (AHFMR). P.S. is a Scientist of the AHFMR. The Julia McFarlane Diabetes Research Centre is supported by the Diabetes Association (Foothills).

REFERENCES

Acha-Orbea, H., and McDevitt, H. O. (1987). The first external domain of the nonobese diabetic mouse class II I-A beta chain is unique. *Proc. Natl. Acad. Sci. USA* **84,** 2435–2439.

Achenbach, P., Bonifacio, E., Williams, A. J., Ziegler, A. G., Gale, E. A., and Bingley, P. J. (2008). Autoantibodies to IA-2beta improve diabetes risk assessment in high-risk relatives. *Diabetologia* **51,** 488–492.

Aichele, P., Kyburz, D., Ohashi, P. S., Odermatt, B., Zinkernagel, R. M., Hengartner, H., and Pircher, H. (1994). Peptide-induced T-cell tolerance to prevent autoimmune diabetes in a transgenic mouse model. *Proc. Natl. Acad. Sci. USA* **91,** 444–448.

Albert, M. L., Sauter, B., and Bhardwaj, N. (1998). Dendritic cells acquire antigen from apoptotic cells and induce class I-restricted CTLs. *Nature* **392,** 86–89.

Allison, J., and Strasser, A. (1998). Mechanisms of beta cell death in diabetes: A minor role for CD95. *Proc. Natl. Acad. Sci. USA* **95,** 13818–13822.

Allison, J., Thomas, H. E., Catterall, T., Kay, T. W., and Strasser, A. (2005). Transgenic expression of dominant-negative Fas-associated death domain protein in beta cells protects against Fas ligand-induced apoptosis and reduces spontaneous diabetes in nonobese diabetic mice. *J. Immunol.* **175,** 293–301.

Amrani, A., Verdaguer, J., Anderson, B., Utsugi, T., Bou, S., and Santamaria, P. (1999). Perforin-independent beta-cell destruction by diabetogenic CD8(+) T lymphocytes in transgenic nonobese diabetic mice. *J. Clin. Invest.* **103,** 1201–1209.

Amrani, A., Verdaguer, J., Serra, P., Tafuro, S., Tan, R., and Santamaria, P. (2000a). Progression of autoimmune diabetes driven by avidity maturation of a T-cell population. *Nature* **406,** 739–742.

Amrani, A., Verdaguer, J., Thiessen, S., Bou, S., and Santamaria, P. (2000b). IL-1alpha, IL-1beta, and IFN-gamma mark beta cells for Fas-dependent destruction by diabetogenic CD4(+) T lymphocytes. *J. Clin. Invest.* **105,** 459–468.

Amrani, A., Serra, P., Yamanouchi, J., Han, B., Thiessen, S., Verdaguer, J., and Santamaria, P. (2002). CD154-dependent priming of diabetogenic CD4(+) T cells dissociated from activation of antigen-presenting cells. *Immunity* **16,** 719–732.

Anderson, M. S., and Bluestone, J. A. (2005). The NOD mouse: A model of immune dysregulation. *Annu. Rev. Immunol.* **23,** 447–485.

Ansari, M. J., Salama, A. D., Chitnis, T., Smith, R. N., Yagita, H., Akiba, H., Yamazaki, T., Azuma, M., Iwai, H., Khoury, S. J., Auchincloss, H., Jr., and Sayegh, M. H. (2003). The programmed death-1 (PD-1) pathway regulates autoimmune diabetes in nonobese diabetic (NOD) mice. *J. Exp. Med.* **198,** 63–69.

Arden, S. D., Zahn, T., Steegers, S., Webb, S., Bergman, B., O'Brien, R. M., and Hutton, J. C. (1999). Molecular cloning of a pancreatic islet-specific glucose-6-phosphatase catalytic subunit-related protein. *Diabetes* **48,** 531–542.

Baekkeskov, S., Aanstoot, H. J., Christgau, S., Reetz, A., Solimena, M., Cascalho, M., Folli, F., Richter-Olesen, H., and De Camilli, P. (1990). Identification of the 64K autoantigen in insulin-dependent diabetes as the GABA-synthesizing enzyme glutamic acid decarboxylase. *Nature* **347,** 151–156.

Baker, C., Petrich de Marquesini, L. G., Bishop, A. J., Hedges, A. J., Dayan, C. M., and Wong, F. S. (2008). Human CD8 responses to a complete epitope set from preproinsulin: Implications for approaches to epitope discovery. *J. Clin. Immunol.* 350–360.

Bendelac, A., Carnaud, C., Boitard, C., and Bach, J. F. (1987). Syngeneic transfer of autoimmune diabetes from diabetic NOD mice to healthy neonates. Requirement for both L3T4+ and Lyt-2+ T cells. *J. Exp. Med.* **166,** 823–832.

Bennett, S. R., Carbone, F. R., Karamalis, F., Flavell, R. A., Miller, J. F., and Heath, W. R. (1998). Help for cytotoxic-T-cell responses is mediated by CD40 signalling. *Nature* **393,** 478–480.

Bercovici, N., Heurtier, A., Vizler, C., Pardigon, N., Cambouris, C., Desreumaux, P., and Liblau, R. (2000). Systemic administration of agonist peptide blocks the progression of spontaneous CD8-mediated autoimmune diabetes in transgenic mice without bystander damage. *J. Immunol.* **165,** 202–210.

Blanas, E., Carbone, F. R., Allison, J., Miller, J. F., and Heath, W. R. (1996). Induction of autoimmune diabetes by oral administration of autoantigen. *Science* **274,** 1707–1709.

Blancou, P., Mallone, R., Martinuzzi, E., Severe, S., Pogu, S., Novelli, G., Bruno, G., Charbonnel, B., Dolz, M., Chaillous, L., van Endert, P., and Bach, J. M. (2007). Immunization of HLA class I transgenic mice identifies autoantigenic epitopes eliciting dominant responses in type 1 diabetes patients. *J. Immunol.* **178,** 7458–7466.

Bouneaud, C., Kourilsky, P., and Bousso, P. (2000). Impact of negative selection on the T cell repertoire reactive to a self-peptide: A large fraction of T cell clones escapes clonal deletion. *Immunity* **13,** 829–840.

Bowie, L., Tite, J., and Cooke, A. (1999). Generation and maintenance of autoantigen-specific CD8(+) T cell clones isolated from NOD mice. *J. Immunol. Methods* **228,** 87–95.

Bresson, D., Togher, L., Rodrigo, E., Chen, Y., Bluestone, J. A., Herold, K. C., and von Herrath, M. (2006). Anti-CD3 and nasal proinsulin combination therapy enhances remission from recent-onset autoimmune diabetes by inducing Tregs. *J. Clin. Invest.* **116,** 1371–1381.

Burton, A. R., Vincent, E., Lennon, G. P., Smeltzer, M., Li, C. S., Haskins, K., Hutton, J., Tisch, R. M., Sercarz, E. E., Santamaria, P., Workman, C. J., and Vignali, D. A. (2008). On the Pathogenicity of Autoantigen-Specific T Cell Receptors. *Diabetes* 1321–1330.

Busick, R. Y., Aguilera, C., and Quinn, A. (2007). Dominant CTL-inducing epitopes on GAD65 are adjacent to or overlap with dominant Th-inducing epitopes. *Clin. Immunol.* **122,** 298–311.

Chen, Y. G., Choisy-Rossi, C. M., Holl, T. M., Chapman, H. D., Besra, G. S., Porcelli, S. A., Shaffer, D. J., Roopenian, D., Wilson, S. B., and Serreze, D. V. (2005). Activated NKT cells inhibit autoimmune diabetes through tolerogenic recruitment of dendritic cells to pancreatic lymph nodes. *J. Immunol.* **174,** 1196–1204.

Choisy-Rossi, C. M., Holl, T. M., Pierce, M. A., Chapman, H. D., and Serreze, D. V. (2004). Enhanced pathogenicity of diabetogenic T cells escaping a non-MHC gene-controlled near death experience. *J. Immunol.* **173,** 3791–3800.

Chong, M. M., Chen, Y., Darwiche, R., Dudek, N. L., Irawaty, W., Santamaria, P., Allison, J., Kay, T. W., and Thomas, H. E. (2004). Suppressor of cytokine signaling-1 overexpression protects pancreatic beta cells from CD8+ T cell-mediated autoimmune destruction. *J. Immunol.* **172,** 5714–5721.

Christen, U., McGavern, D. B., Luster, A. D., von Herrath, M. G., and Oldstone, M. B. (2003). Among CXCR3 chemokines, IFN-gamma-inducible protein of 10 kDa (CXC chemokine ligand (CXCL) 10) but not monokine induced by IFN-gamma (CXCL9) imprints a pattern for the subsequent development of autoimmune disease. *J. Immunol.* **171,** 6838–6345.

Christianson, S. W., Shultz, L. D., and Leiter, E. H. (1993). Adoptive transfer of diabetes into immunodeficient NOD-scid/scid mice. Relative contributions of CD4+ and CD8+ T-cells from diabetic versus prediabetic NOD.NON-Thy-1a donors. *Diabetes* **42**, 44–55.

Cram, D. S., Faulkner-Jones, B., Kun, J., and Harrison, L. C. (1995). Glutamic acid decarboxylase-67 (GAD67): Expression relative to GAD65 in human islets and mapping of autoantibody epitopes. *Endocrinology* **136**, 1111–1119.

Darwiche, R., Chong, M. M., Santamaria, P., Thomas, H. E., and Kay, T. W. (2003). Fas is detectable on beta cells in accelerated, but not spontaneous, diabetes in nonobese diabetic mice.. *J. Immunol.* **170**, 6292–6297.

DiLorenzo, T. P., and Serreze, D. V. (2005). The good turned ugly: Immunopathogenic basis for diabetogenic CD8+ T cells in NOD mice. *Immunol. Rev.* **204**, 250–263.

DiLorenzo, T. P., Graser, R. T., Ono, T., Christianson, G. J., Chapman, H. D., Roopenian, D. C., Nathenson, S. G., and Serreze, D. V. (1998). Major histocompatibility complex class I-restricted T cells are required for all but the end stages of diabetes development in nonobese diabetic mice and use a prevalent T cell receptor alpha chain gene rearrangement. *Proc. Natl. Acad. Sci. USA* **95**, 12538–12543.

Dudda, J. C., and Martin, S. F. (2004). Tissue targeting of T cells by DCs and microenvironments. *Trends Immunol.* **25**, 417–421.

Dudek, N. L., Thomas, H. E., Mariana, L., Sutherland, R. M., Allison, J., Estella, E., Angstetra, E., Trapani, J. A., Santamaria, P., Lew, A. M., and Kay, T. W. (2006). Cytotoxic T-cells from T-cell receptor transgenic NOD8.3 mice destroy beta-cells via the perforin and Fas pathways. *Diabetes* **55**, 2412–2418.

Ejrnaes, M., Videbaek, N., Christen, U., Cooke, A., Michelsen, B. K., and von Herrath, M. (2005). Different diabetogenic potential of autoaggressive CD8+ clones associated with IFN-gamma-inducible protein 10 (CXC chemokine ligand 10) production but not cytokine expression, cytolytic activity, or homing characteristics. *J. Immunol.* **174**, 2746–2755.

Enee, E., Martinuzzi, E., Blancou, P., Bach, J. M., Mallone, R., and van Endert, P. (2008). Equivalent specificity of peripheral blood and islet-infiltrating CD8+ T lymphocytes in spontaneously diabetic HLA-A2 transgenic NOD mice. *J. Immunol.* **180**, 5430–5438.

Fife, B. T., Guleria, I., Gubbels Bupp, M., Eagar, T. N., Tang, Q., Bour-Jordan, H., Yagita, H., Azuma, M., Sayegh, M. H., and Bluestone, J. A. (2005). Insulin-induced remission in new-onset NOD mice is maintained by the PD-1-PD-L1 pathway. *J. Exp. Med.* **203**, 2737–2747.

Filippi, C., Bresson, D., and von Herrath, M. (2005). Antigen-specific induction of regulatory T cells for type 1 diabetes therapy. *Int. Rev. Immunol.* **24**, 341–360.

French, M. B., Allison, J., Cram, D. S., Thomas, H. E., Dempsey-Collier, M., Silva, A., Georgiou, H. M., Kay, T. W., Harrison, L. C., and Lew, A. M. (1997). Transgenic expression of mouse proinsulin II prevents diabetes in nonobese diabetic mice. *Diabetes* **46**, 34–39.

Frigerio, S., Junt, T., Lu, B., Gerard, C., Zumsteg, U., Hollander, G. A., and Piali, L. (2002). Beta cells are responsible for CXCR3-mediated T-cell infiltration in insulitis. *Nat. Med.* **8**, 1414–1420.

Gagnerault, M. C., Luan, J. J., Lotton, C., and Lepault, F. (2002). Pancreatic lymph nodes are required for priming of beta cell reactive T cells in NOD mice. *J. Exp. Med.* **196**, 369–377.

Gao, W., Demirci, G., Strom, T. B., and Li, X. C. (2003). Stimulating PD-1-negative signals concurrent with blocking CD154 co-stimulation induces long-term islet allograft survival. *Transplantation* **76**, 994–999.

Graser, R. T., DiLorenzo, T. P., Wang, F., Christianson, G. J., Chapman, H. D., Roopenian, D. C., Nathenson, S. G., and Serreze, D V. (2000). Identification of a CD8 T cell that can independently mediate autoimmune diabetes development in the complete absence of CD4 T cell helper functions. *J. Immunol.* **164**, 3913–3918.

Green, E. A., Eynon, E. E., and Flavell, R. A. (1998). Local expression of TNFalpha in neonatal NOD mice promotes diabetes by enhancing presentation of islet antigens. *Immunity* **9**, 733–743.

Green, E. A., Wong, F. S., Eshima, K., Mora, C., and Flavell, R. A. (2000). Neonatal tumor necrosis factor alpha promotes diabetes in nonobese diabetic mice by CD154-independent antigen presentation to CD8(+) T cells. *J. Exp. Med.* **191,** 225–238.
Gregori, S., Giarratana, N., Smiroldo, S., and Adorini, L. (2003). Dynamics of pathogenic and suppressor T cells in autoimmune diabetes development. *J. Immunol.* **171,** 4040–4047.
Grewal, I. S., and Flavell, R. A. (1998). CD40 and CD154 in cell-mediated immunity. *Annu. Rev. Immunol.* **16,** 111–135.
Griseri, T., Beaudoin, L., Novak, J., Mars, L. T., Lepault, F., Liblau, R., and Lehuen, A. (2005). Invariant NKT cells exacerbate type 1 diabetes induced by CD8 T cells. *J. Immunol.* **175,** 2091–2101.
Guleria, I., Gubbels Bupp, M., Dada, S., Fife, B., Tang, Q., Ansari, M. J., Trikudanathan, S., Vadivel, N., Fiorina, P., Yagita, H., Azuma, M., Atkinson, M., *et al.* (2007). Mechanisms of PDL1-mediated regulation of autoimmune diabetes. *Clin. Immunol.* **125,** 16–25.
Hamilton-Williams, E. E., Palmer, S. E., Charlton, B., and Slattery, R. M. (2003). Beta cell MHC class I is a late requirement for diabetes. *Proc. Natl. Acad. Sci. USA* **100,** 6688–6693.
Han, B., Serra, P., Amrani, A., Yamanouchi, J., Maree, A. F., Edelstein-Keshet, L., and Santamaria, P. (2005a). Prevention of diabetes by manipulation of anti-IGRP autoimmunity: High efficiency of a low-affinity peptide. *Nat. Med.* **11,** 645–652.
Han, B., Serra, P., Yamanouchi, J., Amrani, A., Elliott, J. F., Dickie, P., Dilorenzo, T. P., and Santamaria, P. (2005b). Developmental control of CD8 T cell-avidity maturation in autoimmune diabetes. *J. Clin. Invest.* **115,** 1879–1887.
Hanayama, R., Tanaka, M., Miyasaka, K., Aozasa, K., Koike, M., Uchiyama, Y., and Nagata, S. (2004). Autoimmune disease and impaired uptake of apoptotic cells in MFG-E8-deficient mice. *Science* **304,** 1147–1150.
Hanninen, A., Jalkanen, S., Salmi, M., Toikkanen, S., Nikolakaros, G., and Simell, O. (1992). Macrophages, T cell receptor usage, and endothelial cell activation in the pancreas at the onset of insulin-dependent diabetes mellitus. *J. Clin. Invest.* **90,** 1901–1910.
Hanninen, A., Nurmela, R., Maksimow, M., Heino, J., Jalkanen, S., and Kurts, C. (2007). Islet beta-cell-specific T cells can use different homing mechanisms to infiltrate and destroy pancreatic islets. *Am. J. Pathol.* **170,** 240–250.
Hartemann-Heurtier, A., Mars, L. T., Bercovici, N., Desbois, S., Cambouris, C., Piaggio, E., Zappulla, J., Saoudi, A., and Liblau, R. S. (2004). An altered self-peptide with superagonist activity blocks a CD8-mediated mouse model of type 1 diabetes. *J. Immunol.* **172,** 915–922.
Hassainya, Y., Garcia-Pons, F., Kratzer, R., Lindo, V., Greer, F., Lemonnier, F. A., Niedermann, G., and van Endert, P. M. (2005). Identification of naturally processed HLA-A2-restricted proinsulin epitopes by reverse immunology. *Diabetes* **54,** 2053–2059.
Heath, W. R., and Carbone, F. R. (2001). Cross-presentation, dendritic cells, tolerance and immunity. *Annu. Rev. Immunol.* **19,** 47–64.
Henquin, J. C., Nenquin, M., Szollosi, A., Kubosaki, A., and Louis Notkins, A. (2008). Insulin secretion in islets from mice with a double knockout for the dense core vesicle proteins islet antigen-2 (IA-2) and IA-2beta. *J. Endocrinol.* **196,** 573–581.
Herrera, P. L., Harlan, D. M., and Vassalli, P. (2000). A mouse CD8 T cell-mediated acute autoimmune diabetes independent of the perforin and Fas cytotoxic pathways: Possible role of membrane TNF. *Proc. Natl. Acad. Sci. USA* **97,** 279–284.
Hoglund, P., Mintern, J., Waltzinger, C., Heath, W., Benoist, C., and Mathis, D. (1999). Initiation of autoimmune diabetes by developmentally regulated presentation of islet cell antigens in the pancreatic lymph nodes. *J. Exp. Med.* **189,** 331–339.
Howson, J. M., Dunger, D. B., Nutland, S., Stevens, H., Wicker, L. S., and Todd, J. A. (2007). A type 1 diabetes subgroup with a female bias is characterised by failure in tolerance to thyroid peroxidase at an early age and a strong association with the cytotoxic T-lymphocyte-associated antigen-4 gene. *Diabetologia* **50,** 741–746.

Itoh, N., Hanafusa, T., Miyazaki, A., Miyagawa, J., Yamagata, K., Yamamoto, K., Waguri, M., Imagawa, A., Tamura, S., Inada, M., Kawata, S., Tarui, S., Kono, K., and Matsuzawa, Y. (1993). Mononuclear cell infiltration and its relation to the expression of major histocompatibility complex antigens and adhesion molecules in pancreas biopsy specimens from newly diagnosed insulin-dependent diabetes mellitus patients. *J. Clin. Invest.* **92,** 2313–2322.

Itoh, N., Imagawa, A., Hanafusa, T., Waguri, M., Yamamoto, K., Iwahashi, H., Moriwaki, M., Nakajima, H., Miyagawa, J., Namba, M., Makino, S., Nagata, S., *et al.* (1997). Requirement of Fas for the development of autoimmune diabetes in nonobese diabetic mice. *J. Exp. Med.* **186,** 613–618.

Jaeckel, E., Lipes, M. A., and von Boehmer, H. (2004). Recessive tolerance to preproinsulin 2 reduces but does not abolish type 1 diabetes. *Nat. Immunol.* **5,** 1028–1035.

Jarchum, I., Baker, J. C., Yamada, T., Takaki, T., Marron, M. P., Serreze, D. V., and DiLorenzo, T. P. (2007). In vivo cytotoxicity of insulin-specific CD8+ T-cells in HLA-A*0201 transgenic NOD mice. *Diabetes* **56,** 2551–2560.

Jarchum, I., Nichol, L., Trucco, M., Santamaria, P., and Dilorenzo, T. P. (2008). Identification of novel IGRP epitopes targeted in type 1 diabetes patients. *Clin. Immunol.* 359–365.

de Jersey, J., Snelgrove, S. L., Palmer, S. E., Teteris, S. A., Mullbacher, A., Miller, J. F., and Slattery, R. M. (2007). Beta cells cannot directly prime diabetogenic CD8 T cells in nonobese diabetic mice. *Proc. Natl. Acad. Sci. USA* **104,** 1295–1300.

Kagi, D., Odermatt, B., Ohashi, P. S., Zinkernagel, R. M., and Hengartner, H. (1996). Development of insulitis without diabetes in transgenic mice lacking perforin-dependent cytotoxicity. *J. Exp. Med.* **183,** 2143–2152.

Kagi, D., Odermatt, B., Seiler, P., Zinkernagel, R. M., Mak, T. W., and Hengartner, H. (1997). Reduced incidence and delayed onset of diabetes in perforin-deficient nonobese diabetic mice. *J. Exp. Med.* **186,** 989–997.

Kanagawa, O., Shimizu, J., and Vaupel, B. A. (2000). Thymic and postthymic regulation of diabetogenic CD8 T cell development in TCR transgenic nonobese diabetic (NOD) mice. *J. Immunol.* **164,** 5466–5473.

Katz, J. D., Wang, B., Haskins, K., Benoist, C., and Mathis, D. (1993). Following a diabetogenic T cell from genesis through pathogenesis. *Cell* **74,** 1089–1100.

Keir, M. E., Liang, S. C., Guleria, I., Latchman, Y. E., Qipo, A., Albacker, L. A., Koulmanda, M., Freeman, G. J., Sayegh, M. H., and Sharpe, A. H. (2006). Tissue expression of PD-L1 mediates peripheral T cell tolerance. *J. Exp. Med.* **203,** 883–895.

Keir, M. E., Freeman, G. J., and Sharpe, A. H. (2007). PD-1 regulates self-reactive CD8+ T cell responses to antigen in lymph nodes and tissues. *J. Immunol.* **179,** 5064–5070.

Keir, M. E., Butte, M. J., Freeman, G. J., and Sharpe, A. H. (2008). PD-1 and its ligands in tolerance and immunity. *Annu. Rev. Immunol.* **26,** 677–704.

Kesmir, C., Nussbaum, A. K., Schild, H., Detours, V., and Brunak, S. (2002). Prediction of proteasome cleavage motifs by neural networks. *Protein Eng.* **15,** 287–296.

Kessler, B., Hudrisier, D., Schroeter, M., Tschopp, J., Cerottini, J. C., and Luescher, I. F. (1998). Peptide modification or blocking of CD8, resulting in weak TCR signaling, can activate CTL for Fas- but not perforin-dependent cytotoxicity or cytokine production. *J Immunol.* **161,** 6939–6946.

Kim, J., Richter, W., Aanstoot, H. J., Shi, Y., Fu, Q., Rajotte, R., Warnock, G., and Baekkeskov, S. (1993). Differential expression of GAD65 and GAD67 in human, rat, and mouse pancreatic islets. *Diabetes* **42,** 1799–1808.

Kim, S., Kim, K. A., Hwang, D. Y., Lee, T. H., Kayagaki, N., Yagita, H., and Lee, M. S. (2000). Inhibition of autoimmune diabetes by Fas ligand: The paradox is solved. *J. Immunol.* **164,** 2931–2936.

Kishimoto, H., and Sprent, J. (2001). A defect in central tolerance in NOD mice. *Nat. Immunol.* **2,** 1025–1031.

Kreuwel, H. T., Morgan, D. J., Krahl, T., Ko, A., Sarvetnick, N., and Sherman, L. A. (1999). Comparing the relative role of perforin/granzyme versus Fas/Fas ligand cytotoxic pathways in CD8+ T cell-mediated insulin-dependent diabetes mellitus. *J. Immunol.* **163,** 4335–4341.

Krishnamurthy, B., Dudek, N. L., McKenzie, M. D., Purcell, A. W., Brooks, A. G., Gellert, S., Colman, P. G., Harrison, L. C., Lew, A. M., Thomas, H. E., and Kay, T. W. (2006). Responses against islet antigens in NOD mice are prevented by tolerance to proinsulin but not IGRP. *J. Clin. Invest.* **116,** 3258–3265.

Krishnamurthy, B., Mariana, L., Gellert, S. A., Colman, P. G., Harrison, L. C., Lew, A. M., Santamaria, P., Thomas, H. E., and Kay, T. W. (2008). Autoimmunity to both proinsulin and IGRP is required for diabetes in nonobese diabetic 8.3 TCR transgenic mice. *J. Immunol.* **180,** 4458–4464.

Kubosaki, A., Gross, S., Miura, J., Saeki, K., Zhu, M., Nakamura, S., Hendriks, W., and Notkins, A. L. (2004a). Targeted disruption of the IA-2beta gene causes glucose intolerance and impairs insulin secretion but does not prevent the development of diabetes in NOD mice. *Diabetes* **53,** 1684–1691.

Kubosaki, A., Miura, J., and Notkins, A. L. (2004b). IA-2 is not required for the development of diabetes in NOD mice. *Diabetologia* **47,** 149–150.

Kurts, C., Carbone, F. R., Barnden, M., Blanas, E., Allison, J., Heath, W. R., and Miller, J. F. (1997). CD4+ T cell help impairs CD8+ T cell deletion induced by cross-presentation of self-antigens and favors autoimmunity. *J. Exp. Med.* **186,** 2057–2062.

Kurts, C., Miller, J. F., Subramaniam, R. M., Carbone, F. R., and Heath, W. R. (1998). Major histocompatibility complex class I-restricted cross-presentation is biased towards high dose antigens and those released during cellular destruction. *J. Exp. Med.* **188,** 409–414.

Kyburz, D., Aichele, P., Speiser, D. E., Hengartner, H., Zinkernagel, R. M., and Pircher, H. (1993). T cell immunity after a viral infection versus T cell tolerance induced by soluble viral peptides. *Eur. J. Immunol.* **23,** 1956–1962.

Lan, M. S., Wasserfall, C., Maclaren, N. K., and Notkins, A. L. (1996). IA-2, a transmembrane protein of the protein tyrosine phosphatase family, is a major autoantigen in insulin-dependent diabetes mellitus. *Proc. Natl. Acad. Sci. USA* **93,** 6367–6370.

Lang, K. S., Recher, M., Junt, T., Navarini, A. A., Harris, N. L., Freigang, S., Odermatt, B., Conrad, C., Ittner, L. M., Bauer, S., Luther, S. A., Uematsu, S., *et al.* (2005). Toll-like receptor engagement converts T-cell autoreactivity into overt autoimmune disease. *Nat. Med.* **11,** 138–145.

Langholz, B., Tuomilehto-Wolf, E., Thomas, D., Pitkaniemi, J., and Tuomilehto, J. (1995). Variation in HLA-associated risks of childhood insulin-dependent diabetes in the Finnish population: I. Allele effects at A, B, and DR loci. DiMe Study Group. Childhood Diabetes in Finland. *Genet Epidemiol.* **12,** 441–453.

Latchman, Y. E., Liang, S. C., Wu, Y., Chernova, T., Sobel, R. A., Klemm, M., Kuchroo, V. K., Freeman, G. J., and Sharpe, A. H. (2004). PD-L1-deficient mice show that PD-L1 on T cells, antigen-presenting cells, and host tissues negatively regulates T cells. *Proc. Natl. Acad. Sci. USA* **101,** 10691–10696.

Lieberman, S. M., Evans, A. M., Han, B., Takaki, T., Vinnitskaya, Y., Caldwell, J. A., Serreze, D. V., Shabanowitz, J., Hunt, D. F., Natheson, S. G., Santamaria, P., and DiLorenzo, T. P. (2003). Identification of the beta cell antigen targeted by a prevalent population of pathogenic CD8+ T cells in autoimmune diabetes. *Proc. Natl. Acad. Sci. USA* **100,** 8384–8388.

Lieberman, S. M., Takaki, T., Han, B., Santamaria, P., Serreze, D. V., and DiLorenzo, T. P. (2004). Individual nonobese diabetic mice exhibit unique patterns of CD8+ T cell reactivity to three islet antigens, including the newly identified widely expressed dystrophia myotonica kinase. *J. Immunol.* **173,** 6727–6734.

Liston, A., Lesage, S., Gray, D. H., O'Reilly, L. A., Strasser, A., Fahrer, A. M., Boyd, R. L., Wilson, J., Baxter, A. G., Gallo, E. M., Crabtree, G. R., Peng, K., et al. (2004). Generalized resistance to thymic deletion in the NOD mouse; a polygenic trait characterized by defective induction of Bim. *Immunity* **21,** 817–830.

Liston, A., Hardy, K., Pittelkow, Y., Wilson, S. R., Makaroff, L. E., Fahrer, A. M., and Goodnow, C. C. (2007). Impairment of organ-specific T cell negative selection by diabetes susceptibility genes: Genomic analysis by mRNA profiling. *Genome Biol.* **8,** R12.

Lo, J., Peng, R. H., Barker, T., Xia, C. Q., and Clare-Salzler, M. J. (2006). Peptide-pulsed immature dendritic cells reduce response to beta cell target antigens and protect NOD recipients from type I diabetes. *Ann. N Y Acad. Sci.* **1079,** 153–156.

Luckashenak, N., Schroeder, S., Endt, K., Schmidt, D., Mahnke, K., Bachmann, M. F., Marconi, P., Deeg, C. A., and Brocker, T. (2008). Constitutive crosspresentation of tissue antigens by dendritic cells controls CD8+ T cell tolerance *in vivo*. *Immunity* **28,** 521–532.

Mallone, R., Martinuzzi, E., Blancou, P., Novelli, G, Afonso, G., Dolz, M., Bruno, G., Chaillous, L., Chatenoud, L., Bach, J. M., and van Endert, P. (2007). CD8+ T-cell responses identify beta-cell autoimmunity in human type 1 diabetes. *Diabetes* **56,** 613–621.

Marron, M. P., Graser, R. T., Chapman, H. D., and Serreze, D. V. (2002). Functional evidence for the mediation of diabetogenic T cell responses by HLA-A2.1 MHC class I molecules through transgenic expression in NOD mice. *Proc. Natl. Acad. Sci. USA* **99,** 13753–13758.

Marshak-Rothstein, A. (2006). Toll-like receptors in systemic autoimmune disease. *Nat. Rev. Immunol.* **6,** 823–835.

Martin, C. C., Bischof, L. J., Bergman, B., Hornbuckle, L. A., Hilliker, C., Frigeri, C., Wahl, D., Svitek, C. A., Wong, R., Goldman, J. K., Oeser, J. K., Lepretre, F., et al. (2001). Cloning and characterization of the human and rat islet-specific glucose-6-phosphatase catalytic subunit-related protein (IGRP) genes. *J. Biol. Chem.* **276,** 25197–25207.

Martinez, N. R., Augstein, P., Moustakas, A. K., Papadopoulos, G. K., Gregori, S., Adorini, L., Jackson, D. C., and Harrison, L. C. (2003). Disabling an integral CTL epitope allows suppression of autoimmune diabetes by intranasal proinsulin peptide. *J. Clin. Invest.* **111,** 1365–1371.

Martinuzzi, E., Novelli, G., Scotto, M., Blancou, P., Bach, J. M., Chaillous, L., Bruno, G., Chatenoud, L., van Endert, P., and Mallone, R. (2008). The frequency and immunodominance of islet-specific CD8+ T-cell responses change after type 1 diabetes diagnosis and treatment. *Diabetes*.

Medarova, Z., Tsai, S., Evgenov, N., Santamaria, P., and Moore, A. (2008). *In vivo* imaging of a diabetogenic CD8+ T cell response during type 1 diabetes progression. *Magn. Reson. Med.* **59,** 712–720.

Millar, D. G., Garza, K. M., Odermatt, B., Elford, A. R., Ono, N., Li, Z., and Ohashi, P. S. (2003). Hsp70 promotes antigen-presenting cell function and converts T-cell tolerance to autoimmunity *in vivo*. *Nat. Med.* **9,** 1469–1476.

Miller, B. J., Appel, M. C., O'Neil, J. J., and Wicker, L. S. (1988). Both the Lyt-2+ and L3T4+ T cell subsets are required for the transfer of diabetes in nonobese diabetic mice. *J. Immunol.* **140,** 52–58.

Moore, A., Grimm, J., Han, B., and Santamaria, P. (2004). Tracking the recruitment of diabetogenic CD8+ T-cells to the pancreas in real time. *Diabetes* **53,** 1459–1466.

Morgan, D. J., Liblau, R., Scott, B., Fleck, S., McDevitt, H. O., Sarvetnick, N., Lo, D., and Sherman, L. A. (1996). CD8(+) T cell-mediated spontaneous diabetes in neonatal mice. *J. Immunol.* **157,** 978–983.

Morgan, D. J., Kreuwel, H. T., and Sherman, L. A. (1999). Antigen concentration and precursor frequency determine the rate of CD8+ T cell tolerance to peripherally expressed antigens. *J. Immunol.* **163,** 723–727.

Moriyama, H., Abiru, N., Paronen, J., Sikora, K., Liu, E., Miao, D., Devendra, D., Beilke, J., Gianani, R., Gill, R. G., and Eisenbarth, G. S. (2003). Evidence for a primary islet autoantigen (preproinsulin 1) for insulitis and diabetes in the nonobese diabetic mouse. *Proc. Natl. Acad. Sci. USA* **100,** 10376–10381.

Mukhopadhaya, A., Hanafusa, T., Jarchum, I., Chen, Y. G., Iwai, Y., Serreze, D. V., Steinman, R. M., Tarbell, K. V., and DiLorenzo, T. P. (2008). Selective delivery of beta cell antigen to dendritic cells *in vivo* leads to deletion and tolerance of autoreactive CD8+ T cells in NOD mice. *Proc. Natl. Acad. Sci. USA* **105,** 6374–6379.

Nagata, M., Santamaria, P., Kawamura, T., Utsugi, T., and Yoon, J. W. (1994). Evidence for the role of CD8+ cytotoxic T cells in the destruction of pancreatic beta-cells in nonobese diabetic mice. *J. Immunol.* **152,** 2042–2050.

Nakayama, M., Abiru, N., Moriyama, H., Babaya, N., Liu, E., Miao, D., Yu, L., Wegmann, D. R., Hutton, J. C., Elliott, J. F., and Eisenbarth, G. S. (2005). Prime role for an insulin epitope in the development of type 1 diabetes in NOD mice. *Nature* **435,** 220–223.

Nejentsev, S., Howson, J. M., Walker, N. M., Szeszko, J., Field, S. F., Stevens, H. E., Reynolds, P., Hardy, M., King, E., Masters, J., Hulme, J., Maier, L. M., *et al.* (2007). Localization of type 1 diabetes susceptibility to the MHC class I genes HLA-B and HLA-A. *Nature* **450,** 887–892.

Nijman, H. W., Houbiers, J. G., Vierboom, M. P., van der Burg, S. H., Drijfhout, J. W., D'Amaro, J., Kenemans, P., Melief, C. J., and Kast, W. M. (1993). Identification of peptide sequences that potentially trigger HLA-A2.1-restricted cytotoxic T lymphocytes. *Eur. J. Immunol.* **23,** 1215–1219.

Nishimura, H., Nose, M., Hiai, H., Minato, N., and Honjo, T. (1999). Development of lupus-like autoimmune diseases by disruption of the PD-1 gene encoding an ITIM motif-carrying immunoreceptor. *Immunity* **11,** 141–151.

Noble, J. A., Valdes, A. M., Bugawan, T. L., Apple, R. J., Thomson, G., and Erlich, H. A. (2002). The HLA class I A locus affects susceptibility to type 1 diabetes. *Hum. Immunol.* **63,** 657–664.

Novak, J., Griseri, T., Beaudoin, L., and Lehuen, A. (2007). Regulation of type 1 diabetes by NKT cells. *Int. Rev. Immunol.* **26,** 49–72.

O'Brien, B. A., Geng, X., Orteu, C. H., Huang, Y., Ghoreishi, M., Zhang, Y., Bush, J. A., Li, G., Finegood, D. T., and Dutz, J. P. (2006). A deficiency in the *in vivo* clearance of apoptotic cells is a feature of the NOD mouse. *J. Autoimmun.* **26,** 104–115.

O'Brien, B. A., Huang, Y., Geng, X., Dutz, J. P., and Finegood, D. T. (2002). Phagocytosis of apoptotic cells by macrophages from NOD mice is reduced. *Diabetes* **51,** 2481–2488.

Ohashi, P. S., Oehen, S., Buerki, K., Pircher, H., Ohashi, C. T., Odermatt, B., Malissen, B., Zinkernagel, R. M., and Hengartner, H. (1991). Ablation of "t;tolerance" and induction of diabetes by virus infection in viral antigen transgenic mice. *Cell* **65,** 305–317.

Oldstone, M. B., Nerenberg, M., Southern, P., Price, J., and Lewicki, H. (1991). Virus infection triggers insulin-dependent diabetes mellitus in a transgenic model: Role of anti-self (virus) immune response. *Cell* **65,** 319–331.

Ouyang, Q., Standifer, N. E., Qin, H., Gottlieb, P., Verchere, C. B., Nepom, G. T., Tan, R., and Panagiotopoulos, C. (2006). Recognition of HLA class I-restricted beta-cell epitopes in type 1 diabetes. *Diabetes* **55,** 3068–3074.

Pakala, S. V., Chivetta, M., Kelly, C. B., and Katz, J. D. (1999). In autoimmune diabetes the transition from benign to pernicious insulitis requires an islet cell response to tumor necrosis factor alpha. *J. Exp. Med.* **189,** 1053–1062.

Panina-Bordignon, P., Lang, R., van Endert, P. M., Benazzi, E., Felix, A. M., Pastore, R. M., Spinas, G. A., and Sinigaglia, F. (1995). Cytotoxic T cells specific for glutamic acid decarboxylase in autoimmune diabetes. *J. Exp. Med.* **181,** 1923–1927.

Parker, K. C., Bednarek, M. A., and Coligan, J. E. (1994). Scheme for ranking potential HLA-A2 binding peptides based on independent binding of individual peptide side-chains. *J. Immunol.* **152,** 163–175.

Perez, V. L., Van Parijs, L., Biuckians, A., Zheng, X. X., Strom, T. B., and Abbas, A. K. (1997). Induction of peripheral T cell tolerance *in vivo* requires CTLA-4 engagement. *Immunity* **6**, 411–417.

Perrin, P. J., Maldonado, J. H., Davis, T. A., June, C. H., and Racke, M. K. (1996). CTLA-4 blockade enhances clinical disease and cytokine production during experimental allergic encephalomyelitis. *J. Immunol.* **157**, 1333–1336.

Petrolonis, A. J., Yang, Q., Tummino, P. J., Fish, S. M., Prack, A. E., Jain, S., Parsons, T. F., Li, P., Dales, N. A., Ge, L., Langston, S. P., Schuller, A. G., *et al.* (2004). Enzymatic characterization of the pancreatic islet-specific glucose-6-phosphatase-related protein (IGRP). *J. Biol. Chem.* **279**, 13976–13983.

Pinkse, G. G., Tysma, O. H., Bergen, C. A., Kester, M. G., Ossendorp, F., van Veelen, P. A., Keymeulen, B., Pipeleers, D., Drijfhout, J. W., and Roep, B. O. (2005). Autoreactive CD8 T cells associated with beta cell destruction in type 1 diabetes. *Proc. Natl. Acad. Sci. USA* **102**, 18425–18430.

Probst, H. C., McCoy, K., Okazaki, T., Honjo, T., and van den Broek, M. (2005). Resting dendritic cells induce peripheral CD8+ T cell tolerance through PD-1 and CTLA-4. *Nat. Immunol.* **6**, 280–286.

Qin, H., Trudeau, J. D., Reid, G. S., Lee, I. F., Dutz, J. P., Santamaria, P., Verchere, C. B., and Tan, R. (2004). Progression of spontaneous autoimmune diabetes is associated with a switch in the killing mechanism used by autoreactive CTL. *Int. Immunol.* **16**, 1657–1662.

Quinn, A., McInerney, M. F., and Sercarz, E. E. (2001). MHC class I-restricted determinants on the glutamic acid decarboxylase 65 molecule induce spontaneous CTL activity. *J. Immunol.* **167**, 1748–1757.

Rammensee, H., Bachmann, J., Emmerich, N. P., Bachor, O. A., and Stevanovic, S. (1999). SYFPEITHI: Database for MHC ligands and peptide motifs. *Immunogenetics* **50**, 213–219.

Robles, D. T., Eisenbarth, G. S., Wang, T., Erlich, H. A., Bugawan, T. L., Babu, S. R., Barriga, K., Norris, J. M., Hoffman, M., Klingensmith, G., Yu, L., and Rewers, M. (2002). Millennium award recipient contribution. Identification of children with early onset and high incidence of anti-islet autoantibodies. *Clin. Immunol.* **102**, 217–224.

Rock, K. L., and Shen, L. (2005). Cross-presentation: Underlying mechanisms and role in immune surveillance. *Immunol. Rev.* **207**, 166–183.

Saeki, K., Zhu, M., Kubosaki, A., Xie, J., Lan, M. S., and Notkins, A. L. (2002). Targeted disruption of the protein tyrosine phosphatase-like molecule IA-2 results in alterations in glucose tolerance tests and insulin secretion. *Diabetes* **51**, 1842–1850.

Salama, A. D., Chitnis, T., Imitola, J., Ansari, M. J., Akiba, H., Tushima, F., Azuma, M., Yagita, H., Sayegh, M. H., and Khoury, S. J. (2003). Critical role of the programmed death-1 (PD-1) pathway in regulation of experimental autoimmune encephalomyelitis. *J. Exp. Med.* **198**, 71–78.

Santamaria, P., Nakhleh, R. E., Sutherland, D. E., and Barbosa, J. J. (1992a). Characterization of T lymphocytes infiltrating human pancreas allograft affected by isletitis and recurrent diabetes. *Diabetes* **41**, 53–61.

Santamaria, P., Noreen, H. J., Lindstrom, A. L., Barbosa, J. J., Faras, A. J., Segall, M., and Rich, S. S. (1992b). DRw52-group haplotypes are frequent acceptors of DRw15-Dw2 DQ genes in DQA1-DRB1 recombination. *Immunogenetics* **36**, 56–63.

Santamaria, P., Utsugi, T., Park, B. J., Averill, N., Kawazu, S., and Yoon, J. W. (1995). Beta-cell-cytotoxic CD8+ T cells from nonobese diabetic mice use highly homologous T cell receptor alpha-chain CDR3 sequences. *J. Immunol.* **154**, 2494–2503.

Savinov, A. Y., Wong, F. S., and Chervonsky, A. V. (2001). IFN-gamma affects homing of diabetogenic T cells. *J. Immunol.* **167**, 6637–6643.

Savinov, A. Y., Tcherepanov, A., Green, E. A., Flavell, R. A., and Chervonsky, A. V. (2003a). Contribution of Fas to diabetes development. *Proc. Natl. Acad. Sci. USA* **100**, 628–632.

Savinov, A. Y., Wong, F. S., Stonebraker, A. C., and Chervonsky, A. V. (2003b). Presentation of antigen by endothelial cells and chemoattraction are required for homing of insulin-specific CD8+ T cells. *J. Exp. Med.* **197,** 643–656.

Schmidt, D., Verdaguer, J., Averill, N., and Santamaria, P. (1997). A mechanism for the major histocompatibility complex-linked resistance to autoimmunity. *J. Exp. Med.* **186,** 1059–1075.

Schmidt, D., Amrani, A., Verdaguer, J., Bou, S., and Santamaria, P. (1999). Autoantigen-independent deletion of diabetogenic CD4+ thymocytes by protective MHC class II molecules. *J. Immunol.* **162,** 4627–4636.

Seewaldt, S., Thomas, H. E., Ejrnaes, M., Christen, U., Wolfe, T., Rodrigo, E., Coon, B., Michelsen, B., Kay, T. W., and von Herrath, M. G. (2000). Virus-induced autoimmune diabetes: most beta-cells die through inflammatory cytokines and not perforin from autoreactive (anti-viral) cytotoxic T-lymphocytes. *Diabetes* **49,** 1801–1809.

Serra, P., Amrani, A., Yamanouchi, J., Han, B., Thiessen, S., Utsugi, T., Verdaguer, J., and Santamaria, P. (2003). CD40 ligation releases immature dendritic cells from the control of regulatory CD4+CD25+ T cells. *Immunity* **19,** 877–889.

Serreze, D. V., Leiter, E. H., Christianson, G. J., Greiner, D., and Roopenian, D. C. (1994). Major histocompatibility complex class I-deficient NOD-B2mnull mice are diabetes and insulitis resistant. *Diabetes* **43,** 505–509.

Serreze, D. V., Chapman, H. D., Varnum, D. S., Gerling, I., Leiter, E. H., and Shultz, L. D. (1997). Initiation of autoimmune diabetes in NOD/Lt mice is MHC class I-dependent. *J. Immunol.* **158,** 3978–3986.

Serreze, D. V., Holl, T. M., Marron, M. P., Graser, R. T., Johnson, E. A., Choisy-Rossi, C., Slattery, R. M., Lieberman, S. M., and DiLorenzo, T. P. (2004). MHC class II molecules play a role in the selection of autoreactive class I-restricted CD8 T cells that are essential contributors to type 1 diabetes development in nonobese diabetic mice. *J. Immunol.* **172,** 871–879.

Serreze, D. V., Choisy-Rossi, C. M., Grier, A. E., Holl, T. M., Chapman, H. D., Gahagan, J. R., Osborne, M. A., Zhang, W., King, B. L., Brown, A., Roopenian, D., and Marron, M. P. (2008). Through regulation of TCR expression levels, an Idd7 region gene(s) interactively contributes to the impaired thymic deletion of autoreactive diabetogenic CD8+ T cells in nonobese diabetic mice. *J. Immunol.* **180,** 3250–3259.

Serreze, D. V., Marron, M. P., and Dilorenzo, T. P. (2007). "Humanized" HLA transgenic NOD mice to identify pancreatic beta cell autoantigens of potential clinical relevance to type 1 diabetes. *Ann. N Y Acad. Sci.* **1103,** 103–111.

Severe, S., Gauvrit, A., Vu, A. T., and Bach, J. M. (2007). CD8+ T lymphocytes specific for glutamic acid decarboxylase 90–98 epitope mediate diabetes in NOD SCID mouse. *Mol. Immunol.* **44,** 2950–2960.

Shameli, A., Yamanouchi, J., Thiessen, S., and Santamaria, P. (2007). Endoplasmic reticulum stress caused by overexpression of islet-specific glucose-6-phosphatase catalytic subunit-related protein in pancreatic Beta-cells. *Rev. Diabet. Stud.* **4,** 25–32.

Shastri, N., Schwab, S., and Serwold, T. (2002). Producing nature's gene-chips: The generation of peptides for display by MHC class I molecules. *Annu. Rev. Immunol.* **20,** 463–493.

Shi, Y., Galusha, S. A., and Rock, K. L. (2006). Cutting edge: Elimination of an endogenous adjuvant reduces the activation of CD8 T lymphocytes to transplanted cells and in an autoimmune diabetes model. *J. Immunol.* **176,** 3905–3908.

Somoza, N., Vargas, F., Roura-Mir, C., Vives-Pi, M., Fernandez-Figueras, M. T., Ariza, A., Gomis, R., Bragado, R., Marti, M., Jaraquemada, D., and Pujol-Borrel, R. (1994). Pancreas in recent onset insulin-dependent diabetes mellitus. Changes in HLA, adhesion molecules and autoantigens, restricted T cell receptor V beta usage, and cytokine profile. *J. Immunol.* **153,** 1360–1377.

Standifer, N. E., Ouyang, Q., Panagiotopoulos, C., Verchere, C. B., Tan, R., Greenbaum, C. J., Pihoker, C., and Nepom, G. T. (2006). Identification of novel HLA-A*0201-restricted

epitopes in recent-onset type 1 diabetic subjects and antibody-positive relatives. *Diabetes* **55,** 3061–3067.
Steinman, R. M., Hawiger, D., and Nussenzweig, M. C. (2003). Tolerogenic dendritic cells. *Annu. Rev. Immunol.* **21,** 685–711.
Sumida, T., Furukawa, M., Sakamoto, A., Namekawa, T., Maeda, T., Zijlstra, M., Iwamoto, I., Koike, T., Yoshida, S., Tomioka, H., et al. (1994). Prevention of insulitis and diabetes in beta 2-microglobulin-deficient non-obese diabetic mice. *Int. Immunol.* **6,** 1445–1449.
Tait, B. D., Colman, P. G., Morahan, G., Marchinovska, L., Dore, E., Gellert, S., Honeyman, M. C., Stephen, K., and Loth, A. (2003). HLA genes associated with autoimmunity and progression to disease in type 1 diabetes. *Tissue Antigens* **61,** 146–153.
Takahashi, K., Honeyman, M. C., and Harrison, L. C. (2001). Cytotoxic T cells to an epitope in the islet autoantigen IA-2 are not disease-specific. *Clin. Immunol.* **99,** 360–364.
Takaki, T., Marron, M. P., Mathews, C. E., Guttmann, S. T., Bottino, R., Trucco, M., DiLorenzo, T. P., and Serreze, D. V. (2006). HLA-A*0201-restricted T cells from humanized NOD mice recognize autoantigens of potential clinical relevance to type 1 diabetes. *J. Immunol.* **176,** 3257–3265.
Tarbell, K. V., Lee, M., Ranheim, E., Chao, C. C., Sanna, M., Kim, S. K., Dickie, P., Teyton, L., Davis, M., and McDevitt, H. (2002). CD4(+) T cells from glutamic acid decarboxylase (GAD)65-specific T cell receptor transgenic mice are not diabetogenic and can delay diabetes transfer. *J. Exp. Med.* **196,** 481–492.
Thebault-Baumont, K., Dubois-Laforgue, D., Krief, F., Briand, J. P., Halbout, P., Vallon-Geoffroy, K., Morin, J., Laloux, V., Lehuen, A., Carel, J. C., Jami, J., Muller, S., et al. (2003). Acceleration of type 1 diabetes mellitus in proinsulin 2-deficient NOD mice. *J. Clin. Invest.* **111,** 851–857.
Thiessen, S., Serra, P., Amrani, A., Verdaguer, J., and Santamaria, P. (2002). T-cell tolerance by dendritic cells and macrophages as a mechanism for the major histocompatibility complex-linked resistance to autoimmune diabetes. *Diabetes* **51,** 325–338.
Thomas, H. E., Parker, J. L., Schreiber, R. D., and Kay, T. W. (1998). IFN-gamma action on pancreatic beta cells causes class I MHC upregulation but not diabetes. *J. Clin. Invest.* **102,** 1249–1257.
Thomas, H. E., Darwiche, R., Corbett, J. A., and Kay, T. W. (1999). Evidence that beta cell death in the nonobese diabetic mouse is Fas independent. *J. Immunol.* **163,** 1562–1569.
Thomas, H. E., Darwiche, R., Corbett, J. A., and Kay, T. W. (2002). Interleukin-1 plus gamma-interferon-induced pancreatic beta-cell dysfunction is mediated by beta-cell nitric oxide production. *Diabetes* **51,** 311–316.
Tisch, R., and McDevitt, H. (1996). Insulin-dependent diabetes mellitus. *Cell* **85,** 291–297.
Tisch, R., Yang, X. D., Singer, S. M., Liblau, R. S., Fugger, L., and McDevitt, H. O. (1993). Immune responses to glutamic acid decarboxylase correlates with insulitis in non-obese diabetic mice. *Nature* **366,** 72.
Tivol, E. A., Borriello, F., Schweitzer, A. N., Lynch, W. P., Bluestone, J. A., and Sharpe, A. H. (1995). *Immunity* **3,** 541.
Todd, J. A., and Wicker, L. S. (2001). Genetic protection from the inflammatory disease type 1 diabetes in humans and animal models. *Immunity* **15,** 387–395.
Todd, J. A., Bell, J. I., and McDevitt, H. O. (1987). HLA-DQ beta gene contributes to susceptibility and resistance to insulin-dependent diabetes mellitus. *Nature* **329,** 599–604.
Toma, A., Haddouk, S., Briand, J. P., Camoin, L., Gahery, H., Connan, F., Dubois-Laforgue, D., Caillat-Zucman, S., Guillet, J. G., Carel, J. C., Muller, S., Choppin, J., et al. (2005). Recognition of a subregion of human proinsulin by class I-restricted T cells in type 1 diabetic patients. *Proc. Natl. Acad. Sci. USA* **102,** 10581–10586.
Trudeau, J. D., Dutz, J. P., Arany, E., Hill, D. J., Fieldus, W. E., and Finegood, D. T. (2000). Neonatal beta-cell apoptosis: A trigger for autoimmune diabetes? *Diabetes* **49,** 1–7.

Trudeau, J. D., Kelly-Smith, C., Verchere, C. B., Elliott, J. F., Dutz, J. P., Finegood, D. T., Santamaria, P., and Tan, R. (2003). Prediction of spontaneous autoimmune diabetes in NOD mice by quantification of autoreactive T cells in peripheral blood. *J. Clin. Invest.* **111,** 217–223.

Tsui, H., Razavi, R., Chan, Y., Yantha, J., and Dosch, H. M. (2007). "Sensing" autoimmunity in type 1 diabetes. *Trends Mol. Med.* **13,** 405–413.

Tsui, H., Chan, Y., Tang, L., Winer, S., Cheung, R. K., Paltser, G., Selvanantham, T., Elford, A. R., Ellis, J. R., Becker, D. J., Ohashi, P. S., and Dosch, H. M. (2008). Targeting of pancreatic glia in type 1 diabetes. *Diabetes* **57,** 918–928.

Turley, S., Poirot, L., Hattori, M., Benoist, C., and Mathis, D. (2003). Physiological beta cell death triggers priming of self-reactive T cells by dendritic cells in a type-1 diabetes model. *J. Exp. Med.* **198,** 1527–1537.

Ueda, H., Howson, J. M., Esposito, L., Heward, J., Snook, H., Chamberlain, G., Rainbow, D. B., Hunter, K. M., Smith, A. N., Di Genova, G., Herr, M. H., Dahlman, I., *et al.* (2003). Association of the T-cell regulatory gene CTLA4 with susceptibility to autoimmune disease. *Nature* **423,** 506–511.

Unger, W. W., Pinkse, G. G., Mulder-van der Kracht, S., van der Slik, A. R., Kester, M. G., Ossendorp, F., Drijfhout, J. W., Serreze, D. V., and Roep, B. O. (2007). Human clonal CD8 autoreactivity to an IGRP islet epitope shared between mice and men. *Ann. N Y Acad. Sci.* **1103,** 192–195.

Utsugi, T., Yoon, J. W., Park, B. J., Imamura, M., Averill, N., Kawazu, S., and Santamaria, P. (1996). Major histocompatibility complex class I-restricted infiltration and destruction of pancreatic islets by NOD mouse-derived beta-cell cytotoxic CD8+ T-cell clones *in vivo*. *Diabetes* **45,** 1121–1131.

Verdaguer, J., Yoon, J. W., Anderson, B., Averill, N., Utsugi, T., Park, B. J., and Santamaria, P. (1996). Acceleration of spontaneous diabetes in TCR-beta-transgenic nonobese diabetic mice by beta-cell cytotoxic CD8+ T cells expressing identical endogenous TCR-alpha chains. *J. Immunol.* **157,** 4726–4735.

Verdaguer, J., Schmidt, D., Amrani, A., Anderson, B., Averill, N., and Santamaria, P. (1997). Spontaneous autoimmune diabetes in monoclonal T cell nonobese diabetic mice. *J. Exp. Med.* **186,** 1663–1676.

Verge, C. F., Gianani, R., Kawasaki, E., Yu, L., Pietropaolo, M., Jackson, R. A., Chase, H. P., and Eisenbarth, G. S. (1996). Prediction of type I diabetes in first-degree relatives using a combination of insulin, GAD, and ICA512bdc/IA-2 autoantibodies. *Diabetes* **45,** 926–933.

Videbaek, N., Harach, S., Phillips, J., Hutchings, P., Ozegbe, P., Michelsen, B. K., and Cooke, A. (2003). An islet-homing NOD CD8+ cytotoxic T cell clone recognizes GAD65 and causes insulitis. *J. Autoimmun.* **20,** 97–109.

Viorritto, I. C., Nikolov, N. P., and Siegel, R. M. (2007). Autoimmunity versus tolerance: Can dying cells tip the balance? *Clin. Immunol.* **122,** 125–134.

Walker, L. S., and Abbas, A. K. (2002). The enemy within: keeping self-reactive T cells at bay in the periphery. *Nat. Rev. Immunol.* **2,** 11–19.

Walter, U., Franzke, A., Sarukhan, A., Zober, C., von Boehmer, H., Buer, J., and Lechner, O. (2000). Monitoring gene expression of TNFR family members by beta-cells during development of autoimmune diabetes. *Eur. J. Immunol.* **30,** 1224–1232.

Wang, B., Gonzalez, A., Benoist, C., and Mathis, D. (1996). The role of CD8+ T cells in the initiation of insulin-dependent diabetes mellitus. *Eur. J. Immunol.* **26,** 1762–1769.

Wang, J., Yoshida, T., Nakaki, F., Hiai, H., Okazaki, T., and Honjo, T. (2005). Establishment of NOD-Pdcd1$^{-/-}$ mice as an efficient animal model of type I diabetes. *Proc. Natl. Acad. Sci. USA* **102,** 11823–11828.

Waterhouse, P., Penninger, J. M., Timms, E., Wakeham, A., Shahinian, A., Lee, K. P., Thompson, C. B., Griesser, H., and Mak, T. W. (1995). Lymphoproliferative disorders with early lethality in mice deficient in Ctla-4. *Science* **270,** 985–988.

Wicker, L. S., Clark, J., Fraser, H. I., Garner, V. E., Gonzalez-Munoz, A., Healy, B., Howlett, S., Hunter, K., Rainbow, D., Rosa, R. L., Smink, L. J., Todd, J. A., *et al.* (2005). Type 1 diabetes genes and pathways shared by humans and NOD mice. *J. Autoimmun.* **25**(Suppl.), 29–33.

Winer, S., Tsui, H., Lau, A., Song, A., Li, X., Cheung, R. K., Sampson, A., Afifiyan, F., Elford, A., Jackowski, G., Becker, D. J., Santamaria, P., *et al.* (2003). Autoimmune islet destruction in spontaneous type 1 diabetes is not beta-cell exclusive. *Nat. Med.* **9**, 198–205.

Wong, F. S., Visintin, I., Wen, L., Flavell, R. A., and Janeway, C. A., Jr. (1996). CD8 T cell clones from young nonobese diabetic (NOD) islets can transfer rapid onset of diabetes in NOD mice in the absence of CD4 cells. *J. Exp. Med.* **183**, 67–76.

Wong, F. S., Karttunen, J., Dumont, C., Wen, L., Visintin, I., Pilip, I. M., Shastri, N., Pamer, E. G., and Janeway, C. A., Jr. (1999). Identification of an MHC class I-restricted autoantigen in type 1 diabetes by screening an organ-specific cDNA library. *Nat. Med.* **5**, 1026–1031.

Wong, C. P., Li, L., Frelinger, J. A., and Tisch, R. (2006). Early autoimmune destruction of islet grafts is associated with a restricted repertoire of IGRP-specific CD8+ T cells in diabetic nonobese diabetic mice. *J. Immunol.* **176**, 1637–1644.

Wong, C. P., Stevens, R., Long, B., Li, L., Wang, Y., Wallet, M. A., Goudy, K. S., Frelinger, J. A., and Tisch, R. (2007). Identical beta cell-specific CD8(−) T cell clonotypes typically reside in both peripheral blood lymphocyte and pancreatic islets. *J. Immunol.* **178**, 1388–1395.

Yagi, H., Matsumoto, M., Kunimoto, K., Kawaguchi, J., Makino, S., and Harada, M. (1992). Analysis of the roles of CD4+ and CD8+ T cells in autoimmune diabetes of NOD mice using transfer to NOD athymic nude mice. *Eur. J. Immunol.* **22**, 2387–2393.

Yamamoto, T., Yamato, E., Tashiro, F., Sato, T., Noso, S., Ikegami, H., Tamura, S., Yanagawa, Y., and Miyazaki, J. I. (2004). Development of autoimmune diabetes in glutamic acid decarboxylase 65 (GAD65) knockout NOD mice. *Diabetologia* **47**, 221–224.

Yamanouchi, J., Verdaguer, J., Han, B., Amrani, A., Serra, P., and Santamaria, P. (2003). Cross-priming of diabetogenic T cells dissociated from CTL-induced shedding of beta cell autoantigens. *J. Immunol.* **171**, 6900–6909.

Yamanouchi, J., Rainbow, D., Serra, P., Howlett, S., Hunter, K., Garner, V. E., Gonzalez-Munoz, A., Clark, J., Veijola, R., Cubbon, R., Chen, S. L., Rosa, R., *et al.* (2007). Interleukin-2 gene variation impairs regulatory T cell function and causes autoimmunity. *Nat. Genet.* **39**, 329–337.

Yang, X. D., Tisch, R., Singer, S. M., Cao, Z. A., Liblau, R. S., Schreiber, R. D., and McDevitt, H. O. (1994). Effect of tumor necrosis factor alpha on insulin-dependent diabetes mellitus in NOD mice. I. The early development of autoimmunity and the diabetogenic process. *J. Exp. Med.* **180**, 995–1004.

Zhang, Y., O'Brien, B., Trudeau, J., Tan, R., Santamaria, P., and Dutz, J. P. (2002). *In situ* beta cell death promotes priming of diabetogenic CD8 T lymphocytes. *J. Immunol.* **168**, 1466–1472.

CHAPTER 5

Dysrulation of T Cell Peripheral Tolerance in Type 1 Diabetes

R. Tisch and **B. Wang**

Contents		
	1. Introduction	125
	2. The Autoimmune Process of T1D	126
	3. Dysregulation of Central T Cell Tolerance in T1D	128
	4. Dysregulation of Peripheral T Cell Tolerance in T1D	130
	4.1. Defects in anergy and deletion	130
	4.2. Defects in immunoregulation	132
	5. Summary	140
	Acknowledgments	141
	References	141

Abstract   Type 1 diabetes (T1D) is a T cell-mediated autoimmune disease in which the insulin producing β cells are destroyed. The breakdown of β cell-specific self-tolerance by T cells involves a number of dysregulated events intrinsic and extrinsic to T cells. Herein, we review the key mechanisms that drive β cell autoimmunity, with an emphasis on events that influence the expansion and differentiation of pathogenic T cells in the periphery.

Key Words: Autoimmunity, Diabetes, Immunoregulation, Tolerance.   © 2008 Elsevier Inc.

Department of Microbiology and Immunology, University of North Carolina at Chapel Hill, Chapel Hill, North Carolina

1. INTRODUCTION

The hallmark of a functional immune system is the capacity to distinguish between foreign antigens expressed by pathogens, and self antigens expressed by the body. The lack of a pathological response to self-antigens is dependent on a number of events that occur centrally and peripherally. Central tolerance is induced at the site of lymphocyte development such as the thymus and bone marrow for T cells and B cells, respectively. On the other hand peripheral tolerance occurs at sites of antigen recognition and processing, and includes secondary lymphoid as well as nonlymphoid tissues. Failure of central and/or peripheral tolerance can lead to increased development and expansion of pathogenic effectors, and the subsequent initiation and progression of autoimmunity.

Type 1 diabetes (T1D) is an autoimmune disease characterized by the destruction of the insulin producing β cells found in the pancreatic islets of Langerhans (Anderson and Bluestone, 2005; Bach, 1994; Eisenbarth, 2004; Tisch and McDevitt, 1996). Based largely on studies employing rodent models of spontaneous T1D, such as the nonobese diabetic (NOD) mouse and the Biobreeding (BB) rat, the primary mediators of β cell destruction are $CD4^+$ and $CD8^+$ T cells (Anderson and Bluestone, 2005; Bach, 1994; Eisenbarth, 2004; Tisch and McDevitt, 1996). These pathogenic effector T cells typically exhibit a type 1 phenotype marked by the production of IFNγ and TNFα. The increased frequency of β cell-specific precursors and the preferential skewing towards type 1 T effectors indicate defects in both central and peripheral tolerance. The goal of this review is to highlight the events involved in the breakdown of β cell-specific tolerance within the T cell compartment, with the focus being on peripheral mechanisms. We will discuss recent findings within the field that define the cellular and molecular events that promote the development of β cell-specific pathogenic T cell effectors.

2. THE AUTOIMMUNE PROCESS OF T1D

Currently, the critical events involved in the initiation and progression of β cell autoimmunity remain ill-defined. In general, β cell autoimmunity is thought to be triggered and/or exacerbated in genetically predisposed individuals encountering an ill-defined environmental challenge(s) (Maier and Wicker, 2005; Onengut-Gumuscu and Concannon, 2006; Sarvetnick, 2000; Todd and Wicker, 2001; Wicker *et al.*, 1995, 2005). Once β cell autoimmunity has been initiated, an asymptomatic phase follows which can be monitored by the presence of serum autoantibodies specific for β cell autoantigens such as insulin, glutamic acid decarboxylase 65

(GAD65), and insulinoma-associated tyrosine phosphatase (IA-2) (Anderson and Bluestone, 2005; Bach, 1994; Eisenbarth, 2004; Tisch and McDevitt, 1996). The onset of frank diabetes typically occurs several years after the triggering event, suggesting that β cell autoimmunity involves a progressive loss of self-tolerance.

More than 20 chromosomal loci, referred to as insulin dependent diabetes *(Idd)* loci, contribute to T1D development in humans, and the NOD mouse (Maier and Wicker, 2005; Onengut-Gumuscu and Concannon, 2006; Todd and Wicker, 2001; Wicker *et al.*, 1995, 2005). The fact that a number of genes are associated with disease susceptibility, argues that dysregulation of multiple pathways of self-tolerance is needed for T1D development. In humans, the strongest association with both susceptibility for and resistance to T1D is linked to genes encoding the HLA class II HLA-DR and -DQ molecules. For instance, DR3-DQ2 (DRB1*0301-DQB1*0201) and DR4-DQ8 (DRB1*0401-DQB1*0302) haplotypes are associated with susceptibility whereas the DQB1*0602 allele is linked to dominant protection (Maier and Wicker, 2005; Todd and Wicker, 2001). These associations reflect a key role for the HLA haplotypes in shaping the repertoire of β cell-specific $CD4^+$ T cells. A similar association with disease susceptibility and the MHC class II molecule IA^{g7} is seen in the NOD mouse (Maier and Wicker, 2005; Todd and Wicker, 2001; Wicker *et al.*, 1995). Additional genes encoding insulin, CTLA-4, the IL-2 receptor α chain (CD25), and the lymphocyte-specific tyrosine phosphatase (LYP) among others, have also been identified as candidates that promote disease susceptibility (Maier and Wicker, 2005; Onengut-Gumuscu and Concannon, 2006; Wicker *et al.*, 2005).

Much of our understanding of the immunopathogenesis of T1D comes from studies employing the NOD mouse (Anderson and Bluestone, 2005; Bach, 1994; Tisch and McDevitt, 1996). β cell autoimmunity progresses in NOD mice in relatively well-defined stages or "checkpoints" (Andre *et al.*, 1996). The first checkpoint is marked by the infiltration of the islets or insulitis in 2–3-week-old NOD mice. Dendritic cells (DC) and macrophages are initially detected, followed then by $CD8^+$ and $CD4^+$ T cells, and B cells. Early in disease, a select few β cell autoantigens such as insulin B chain (Krishnamurthy *et al.*, 2006; Nakayama *et al.*, 2005, 2007), GAD65 (Kaufman *et al.*, 1993; Tisch *et al.*, 1993), and islet-specific glucose-6-phosphatase catalytic subunit-related protein (IGRP) are recognized by T cells (Amrani *et al.*, 2000). Priming of naïve β cell-specific $CD4^+$ (and possibly $CD8^+$) T cells occurs in the draining pancreatic lymph nodes (panLN), after which the T cells traffic to the islets (Hoglund *et al.*, 1999). As β cell autoimmunity progresses "epitope spread" occurs as additional intra- and inter-molecular epitopes are targeted by T cells (Kaufman *et al.*, 1993; Tisch *et al.*, 1993). B cells functioning primarily as APC have been reported to play a key role in driving epitope spread (Tian *et al.*, 2006).

Despite the progression of insulitis, the majority of β cell mass remains intact for a number of weeks. At approximately 12 weeks of age, however, insulitis transitions from a benign to an "aggressive" infiltrate thereby marking the onset of checkpoint 2. Hyperglycemia ensues once 90% of the β cells have been destroyed. The events promoting the progression to checkpoint 2 are poorly understood. However, a temporal decrease in the number and/or function of immunoregulatory T cells (Alard et al., 2006; Brusko et al., 2005; Fox and Danska, 1997; Gregg et al., 2004; Gregori et al., 2003; Herman et al., 2004; Lindley et al., 2005; Pop et al., 2005; Tang et al., 2008; Tritt et al., 2008; Wu et al., 2002; You et al., 2004), increased resistance of T cells to suppression (Gregori et al., 2003; You et al., 2005) coupled with avidity maturation of β cell-specific type 1 effector T cells (Amrani et al., 2000) have been reported to coincide with the onset of checkpoint 2.

3. DYSREGULATION OF CENTRAL T CELL TOLERANCE IN T1D

The T cell repertoire is shaped within the thymus by selection events governed by the affinity of the T cell receptor (TCR) for complexes of self-peptide/MHC molecules (self-pMHC) expressed on the surface of thymic epithelial cells (TEC) and DC (Starr et al., 2003; Venanzi et al., 2004). Thymic negative selection is the predominant mechanism by which the autoreactive TCR repertoire is purged. $CD4^+ CD8^+$ thymocytes binding self-pMHC with increasingly high affinity are clonally deleted via apoptotic inducing events (Starr et al., 2003; Venanzi et al., 2004). Negative selection is mediated by a surprisingly large number of tissue-specific antigens (TSA) including β cell-derived proteins that are expressed by medullary TEC (mTEC) (Anderson et al., 2002; Derbinski et al., 2005; Liston et al., 2003). Expression of genes encoding these TSA is regulated by the recently identified transcription factor AIRE (autoimmune regulatory). Mutations in the *AIRE* gene that alter the transcriptional function of the protein are associated with autoimmune polyendocrinopathy-candidiasis-ectodermal dystrophy (APECED), a syndrome characterized by chronic mucocutaneous candidiasis, hypoparathyroidism, and adrenal insufficiency (Mathis and Benoist, 2007). APECED patients also present with a variety of types of autoimmunity such as thyroiditis, hepatitis, and T1D (Mathis and Benoist, 2007). Similarly, mice deficient in AIRE expression develop T cell-mediated autoimmunity targeting a number of tissues (Anderson et al., 2002; Liston et al., 2003). Thymic DC found in the medulla also contribute to thymocyte negative selection. Here, DC are believed to cross-present TSA derived from mTEC (Starr et al., 2003). However, AIRE expression is reported in DC albeit at low levels, and TSA-expression including β cell-autoantigens such as proinsulin, GAD65, and IA-2 has been detected in both thymic and peripheral DC (Pugliese and Diez, 2002;

Pugliese et al., 2001). Therefore, thymic DC may also be an important direct source of TSA, and possible defects in thymic DC would be predicted to impact negative selection events.

A number of factors contribute to the efficiency of negative selection and the resulting frequency of autoreactive T precursors that enter the periphery. For instance, reports indicate that $CD8^+$ $CD4^+$ thymocytes in the NOD mouse are relatively insensitive to apoptosis partly due to aberrant up-regulation of the proapoptotic protein Bim (Liston et al., 2004; Zucchelli et al., 2005). An elevated threshold for clonal deletion in $CD4^+$ $CD8^+$ thymocytes is expected to reduce the efficiency of negative selection. A similar mechanism is thought to explain the strong association between human T1D and a single-nucleotide polymorphism in the *PTPN22* gene, which encodes LYP (Bottini et al., 2006; Maier and Wicker, 2005; Onengut-Gumuscu and Concannon, 2006). LYP functions in T cells as a negative regulator of TCR signaling by directly dephosphorylating the Src kinases Lck and Fyn, ITAMs of the TCRζ/CD3 complex, and ZAP70 among other molecules (Bottini et al., 2006). The polymorphism in the *PTPN22* gene is associated with increased phosphatase activity which down-regulates TCR signaling events more efficiently (Bottini et al., 2006). It has been proposed that the "hyper-phosphatase" activity of the disease associated LYP isoform further dampens TCR signaling in $CD4^+$ $CD8^+$ thymocytes to increase the activational threshold needed for deletion (Bottini et al., 2006).

The level of expression and presentation of TSA-derived peptides is another factor influencing thymic clonal deletion. As noted above, aberrant TSA expression in the thymus due to mutations in the *AIRE* gene promotes the development of tissue-specific autoimmunity. Noteworthy is that a variable nucleotide tandem repeat (VNTR) found upstream of the insulin gene is associated with T1D susceptibility and resistance in humans (Maier and Wicker, 2005; Onengut-Gumuscu and Concannon, 2006; Wicker et al., 2005). The "long-form" of the VNTR corresponds with protection from diabetes and increased insulin mRNA within the thymus (Maier and Wicker, 2005; Onengut-Gumuscu and Concannon, 2006; Wicker et al., 2005). This correlation suggests that deletion of insulin-specific thymocytes is enhanced with increased thymic insulin expression. Indeed, an inverse relationship is seen between the levels of thymic insulin expression and peripheral insulin-specific T cell reactivity in genetically manipulated NOD mice (Chentoufi and Polychronakos, 2002).

The level of presentation of TSA-derived peptides will also be influenced by the peptide binding properties of the respective MHC molecules and the stability of pMHC complexes. The IA^{g7} allele expressed by NOD mice forms a relatively "weak" pMHC complex (Suri et al., 2008) thereby reducing the number of "effective" pMHC complexes and overall avidity of the TCR–pMHC interaction. Consequently, the efficiency of negative selection of β cell-specific T precursors is further reduced.

4. DYSREGULATION OF PERIPHERAL T CELL TOLERANCE IN T1D

Central tolerance is not absolute and T cells specific for self-antigens are detected in the blood of healthy individuals (Arif et al., 2004). A number of peripheral mechanisms are in place to ensure that activation and expansion of autoreactive T cells are limited, and a pathological response is not induced. Active mechanisms of peripheral T cell tolerance include the induction of anergy and deletion, and immunoregulation. Immunological "ignorance" or "indifference" is a passive form of peripheral tolerance in which TSA are either sequestered or self-pMHC complexes found on "resting" APC are insufficient to initiate TCR signaling and T cell activation. For instance, TCR transgenic mice containing $CD4^+$ or $CD8^+$ T cells specific for viral proteins fail to develop diabetes despite expression of the corresponding viral proteins by β cells. Only when APC are activated and a proinflammatory *milieu* is established by viral infection are T cells stimulated and β cells destroyed (Ohashi et al., 1991; von Herrath et al., 1994).

4.1. Defects in anergy and deletion

It is well established that "two signals" are required for naïve T cells to sustain proliferation, and secrete cytokines. Signal 1 is delivered by the TCR following the binding of pMHC (Davis and van der Merwe, 2006). The magnitude of signal 1 is determined by the binding affinity of the TCR, and the number of pMHC complexes found on the surface of APC (Davis and van der Merwe, 2006). Signal 2 is derived from the costimulatory molecule CD28 upon binding of CD80 and CD86 (Chen, 2004). Other molecules such as CD154 (CD40L), OX40, 4–1BB, and ICOS provide costimulation for activated T cells that is necessary for efficient effector cell differentiation (Chen, 2004). TCR signaling alone typically results in aborted T cell division and induction of activation induced cell death (AICD) due to the lack of up-regulation of prosurvival factors such as IL-2 and Bcl-XL (Hernandez et al., 2001; Noel et al., 1996). The remaining T cells that fail to undergo apoptosis usually are unresponsive to subsequent antigen-stimulation. This state of anergy is an active process involving a number of "anergy factors" such as E3 ubiquitin ligases, which block signaling delivered by the TCR complex, CD28, and the IL-2 receptor (Mueller, 2004). A limited number of studies suggest that defects in peripheral AICD intrinsic to T cells exist in NOD mice. Reduced expression of proapoptotic molecules such as caspase 8 and Fas/FasL, and persistent levels of the anti-apoptotic protein c-FLIP have been reported (Arreaza et al., 2003; Decallonne et al., 2003). However, the general consensus is that defective T cell AICD and/or anergy induction have a minimal role in promoting the breakdown of β cell-specific tolerance.

In addition to positive costimulatory receptors, T cell reactivity is controlled by negative regulatory receptors such as CTLA-4 and PD-1 (Okazaki and Honjo, 2007; Walunas et al., 1996). Analogous to CD28, CTLA-4 binds CD80 and CD86 but with increased affinity (Krummel and Allison, 1995). CTLA-4 attenuates T cell activation by competing for CD28 ligation and recruiting serine/threonine and tyrosine phosphatases to inhibit TCRζ chain phosphorylation, and immune synapse formation (Korman et al., 2006). The importance of CTLA-4-mediated regulation of T cell activation is seen in CTLA-4-deficient mice, which develop massive lymphoproliferation (Tivol et al., 1995). Furthermore, blockade of CTLA-4 enhances the progression of insulitis in young NOD mice (Luhder et al., 1998). Splice variants of the human CTLA-4 gene are associated with a number of autoimmune diseases including T1D (Kristiansen et al., 2000; Ueda et al., 2003). Two major isoforms of CTLA-4 exist; a full-length membrane-bound isoform and a soluble isoform lacking the transmembrane domain (Magistrelli et al., 1999). Reduced expression of the soluble CTLA-4 isoform (sCTLA-4) by T cells correlates with autoimmune disease susceptibility. How levels of sCTLA-4 *in vivo* regulate T cell tolerance is currently unclear (Tivol et al., 1995). sCTLA-4 in human serum inhibits T cell proliferation *in vitro* by binding to CD80 and CD86, and blocking costimulatory signaling needed for T cell activation (Saverino et al., 2007; Ueda et al., 2003). Accordingly, it has been proposed that reduced expression of sCTLA-4 may limit "costimulatory molecule blockade" and enhance the capacity of APC to stimulate autoreactive T cells (Saverino et al., 2007; Ueda et al., 2003).

Mice have an additional CTLA-4 splice variant that is membrane bound but lacks the binding domain for CD80 and CD86. Vijayakrishnan et al. demonstrated that this ligand-independent CTLA-4 (liCTLA-4) isoform inhibits T cell activation by directly dephosphorylating the TCRζ chain (Vijayakrishnan et al., 2004). Furthermore, expression of liCTLA-4 is increased in resting memory T cells (Vijayakrishnan et al., 2004), which generally are more sensitive to stimulation compared to naïve T cells. liCTLA-4 may increase the activational threshold of resting memory T cells to limit the stimulatory capacity of self-pMHC complexes. In NOD mice, expression of liCTLA-4 is reduced relative to diabetes-free congenic NOD lines (Ueda et al., 2003). Consequently, resting memory T cells in NOD mice, may be more sensitive to TCR-mediated activation upon binding of self-pMHC complexes.

PD-1 is expressed on activated T cells and similar to CTLA-4 blocks downstream phosphorylation events induced by TCR and CD28, albeit by a different mechanism (Parry et al., 2005). The latter is mediated by an immunoreceptor tyrosine-based switch motif located in the cytoplasmic tail of PD-1 (Parry et al., 2005). A role for PD-1 in regulating T1D is seen in PD-1-deficient NOD mice in which β cell autoimmunity is exacerbated

(Ansari et al., 2003). In addition, ectopic expression by β cells of a PD-1 ligand, namely PD-L1 blocks the development of diabetes in NOD mice (Keir et al., 2006).

4.2. Defects in immunoregulation

Immunoregulation is a dynamic process in which pro- and anti-inflammatory responses compete resulting in re-establishment of immune homeostasis (e.g., self-tolerance). Several cell types contribute to immunoregulation with T cells being the most dominant effector cell type. However, innate cells such as DC also have a significant role by either directly or indirectly controlling proinflammatory (pathogenic) and/or immunoregulatory T cells.

4.2.1. T cell-mediated immunoregulation

In the early 1990s immunoregulation of autoreactive T cells was viewed in the context of Th1 and Th2 cells (Bach, 1994; Liblau et al., 1995; Tisch and McDevitt, 1996). Progression of tissue-specific autoimmunity such as T1D, was attributed to a functional imbalance between pathogenic Th1 cells and immunoregulatory Th2 cells. Self-tolerance was thought to be controlled by tissue-specific Th2 cells, which prevented the development of pathogenic Th1 cells and promoted further differentiation of Th2 cells in a bystander manner mediated by IL-4. Although an oversimplification, this paradigm holds true today. It is now clear that a number of subsets of immunoregulatory T cells (Treg) contribute to the establishment and maintenance of peripheral tolerance (Shevach, 2006). Treg can be broadly divided into effectors that express the FoxP3 transcription factor and "conventional" effectors.

4.2.1.1. Foxp3$^+$ Treg So-called "natural" Treg (nTreg) are characterized by a potent suppressor function that is established in the thymus upon recognition of self-pMHC (Kang et al., 2007; Sakaguchi, 2005; Tang and Bluestone, 2008). Phenotypically, nTreg are defined by FoxP3 expression and constitutive expression of CD25, although other surface markers such as CD39, CD62L, CTLA-4, GITR, and CD127 are also used (Kang et al., 2007; Sakaguchi, 2005; Tang and Bluestone, 2008). Initially, FoxP3 was viewed as a nTreg "master regulator" since transfer of the *FOXP3* gene into naive CD4$^+$ T cells (e.g., CD4$^+$ CD25$^-$ T cells) induced phenotypic and functional changes typical of nTreg (Fontenot et al., 2003; Hori et al., 2003). Recent studies, however, suggest that the molecular basis for the induction and maintenance of nTreg is more complex, involving a number of "transcriptional signatures" independent of FoxP3 and established by signaling via TGFβ, IL-2 and possibly other receptors (Hill et al., 2007). The importance of nTreg in regulating self-tolerance is evident in mice and humans that lack nTreg due to mutations in *FOXP3*, and which develop severe multi-organ autoimmunity (Gambineri *et al.*, 2003; Khattri *et al.*, 2003; Kim *et al.*, 2007). In addition Chen *et al.* showed that FoxP3-deficient NOD mice exhibit an increased incidence and accelerated onset of diabetes relative to wild-type NOD mice (Chen *et al.*, 2005). Upon antigen stimulation, nTreg exhibit a suppressor function that affects several cell types including naïve and effector $CD4^+$ and $CD8^+$ T cells, B cells, and APC. Depending on the *in vitro* and *in vivo* conditions the suppressor function of nTreg is mediated by (but not limited to): (1) cell-cell contact possibly involving CTLA-4 binding, (2) cytokine "deprivation" by CD25 binding of IL-2, (3) bystander suppression via secretion of anti-inflammatory cytokines such as TGFβ, IL-10 and IL-35, and (4) cytolytic activity due granzyme B and perforin release (Kang *et al.*, 2007; Tang and Bluestone, 2008). This variety in effector mechanisms suggests that distinct subsets of nTreg exist. In addition, $CD4^+$ $CD25^-$ T cells have been shown *in vitro* and *in vivo* to up-regulate FoxP3 expression under the appropriate conditions. For example, $CD4^+$ $CD25^-$ T cells stimulated with antigen in cultures supplemented with IL-2 and TGFβ differentiate into FoxP3-expressing "induced" Treg (iTreg), which closely resemble nTreg both phenotypically and functionally (Chen *et al.*, 2003). Since nTreg and iTreg cannot be phenotypically distinguished from one another, the relative role of these two types of FoxP3-expressing effectors in regulating *in vivo* self-tolerance is unclear.

The primary site at which nTreg control β cell autoimmunity is within the islet infiltrate where the function of established effector T cells is suppressed (Chen *et al.*, 2005). In addition, nTreg may function by blocking the priming of naïve β cell-specific T precursors in the draining panLN, and preventing infiltration of the islets (Pop *et al.*, 2005). Several groups have examined nTreg temporally in NOD mice and the blood of diabetic patients to determine whether defects in nTreg contribute to the progression of T1D. The findings have been somewhat mixed, possibly due to different methods of nTreg detection and isolation, and/or the mouse model systems being employed. Studies have demonstrated altered numbers and/or function of nTreg in both NOD mice and diabetic patients (Alard *et al.*, 2006; Brusko *et al.*, 2005; Fox and Danska, 1997; Gregg *et al.*, 2004; Gregori *et al.*, 2003; Herman *et al.*, 2004; Lindley *et al.*, 2005; Mellanby *et al.*, 2007; Pop *et al.*, 2005; Tang *et al.*, 2008; Tritt *et al.*, 2008; Wu *et al.*, 2002; You *et al.*, 2005), whereas some groups have found no significant changes within the nTreg pool. The majority of reports, however, suggest a scenario in which the suppressor function of nTreg declines with the progression of β cell autoimmunity (Alard *et al.*, 2006; Fox and Danska, 1997; Gregg *et al.*, 2004; Gregori *et al.*, 2003; Herman *et al.*, 2004; Lindley *et al.*, 2005; Pop *et al.*, 2005; Tang *et al.*, 2008; Tritt *et al.*, 2008; Wu *et al.*, 2002; You *et al.*, 2005). For instance, Pop *et al.* showed that the *in vitro* and *in vivo* suppressor function of nTreg significantly wanes in NOD mice in an age dependent manner, and that this defect correlates with reduced levels of FoxP3 expression, and TGFβ production by nTreg (Pop *et al.*, 2005). A similar temporal defect in TGFβ expression and reduced suppressor function by nTreg in NOD mice was reported by Gregg *et al.* (Gregg *et al.*, 2004). A temporal reduction of nTreg suppressor function has also been detected in diabetic patients (Brusko *et al.*, 2005; Lindley *et al.*, 2005).

The basis for the aberrant suppressor function of nTreg in NOD mice and diabetic patients has yet to be defined. Since CTLA-4 is directly involved in nTreg/iTreg-mediated immunoregulation (Kang *et al.*, 2007; Tang and Bluestone, 2008), altered expression of the respective isoforms of CTLA-4 by nTreg/iTreg may influence the suppressor function of these effectors. In humans, polymorphisms within the *IL2RA* (CD25) gene region are associated with T1D (Lowe *et al.*, 2007; Onengut-Gumuscu and Concannon, 2006; Wicker *et al.*, 2005). Dysregulation of signaling by or expression of CD25 would be expected to influence the maintenance and/or effector function of nTreg. For instance, IL-2 is critical for the peripheral maintenance of nTreg, and treatment of mice with anti-IL-2 antibody significantly reduces the *in vivo* frequency of nTreg (Fehervari *et al.*, 2006; Murakami *et al.*, 2002; Setoguchi *et al.*, 2005). One possible scenario in NOD mice involves IL-21, a pro-inflammatory cytokine that is primarily produced by activated T cells (Spolski and Leonard, 2008). The IL-21 receptor is expressed by T cells, nTreg, B cells and NK cells. In the case of T cells, IL-21 provides a co-stimulatory signal for proliferation (Spolski and Leonard, 2008). Noteworthy is that the gene encoding IL-21 (along with the *il2* gene) is located in the *idd3* genetic locus, and increased IL-21 expression by T cells has been reported in NOD mice (King *et al.*, 2004; Yamanouchi *et al.*, 2007). Importantly, various studies have demonstrated that IL-21 down-regulates FoxP3 expression and inhibits nTreg suppressor function (Li and Yee, 2008; Piao *et al.*, 2008). Therefore, increasing levels of IL-21 secreted by expanding β cell-specific T effectors in the panLN or islets could reduce levels of FoxP3 expression, and limit the suppressor function of nTreg as β cell autoimmunity proceeds.

Recent work by Tang *et al.* suggests that dysregulation of IL-2 expression is a contributing factor in the aberrant nTreg activity detected in NOD mice (Tang *et al.*, 2008). Tang *et al.* demonstrated that the survival of nTreg in the islets but not panLN decreases with disease progression, and that this defect is attributed to reduced IL-2 production in NOD mice. As noted above, the gene encoding IL-2 is found in the *idd3* genetic locus, and studies have reported reduced IL-2 secretion by T cells in NOD versus NOD mice congenic for the *idd3* interval derived from C57BL/6 (NOD.*idd3*^{B6}) mice (Yamanouchi *et al.*, 2007). NOD.*idd3*^{B6} congenic mice exhibit a reduced incidence and delayed onset of diabetes relative to wild-type NOD mice (Yamanouchi et al., 2007). Tang et al. propose that in NOD mice the survival of islet infiltrating nTreg is impaired due to insufficient levels of IL-2 and the lack of up-regulation of the pro-survival protein Bcl-2. Indeed, nTreg numbers are "normalized" and NOD mice remain diabetes-free with injection of IL-2-antibody complexes (Tang et al., 2008).

Reduced IL-2 and increased IL-21 levels in NOD mice may also impair the induction of iTreg. IL-2 has been shown along with TGFβ to induce FoxP3-expression and nTreg-like suppressor function in $CD25^-$ $CD4^+$ T cells *in vitro* (Chen et al., 2003; Davidson et al., 2007). On the other hand, IL-21 blocks the induction of FoxP3 expression and iTreg differentiation by $CD4^+$ $CD25^-$ T cells *in vitro* (Korn et al., 2007). It is possible that the dysregulated levels of both IL-2 and IL-21 detected in NOD mice limit *in vivo* induction of iTreg, thereby further diminishing the overall pool of $FoxP3^+$ Treg.

An additional factor contributing to insufficient nTreg/iTreg activity in NOD mice is the sensitivity of effector T cells to immunoregulation (Gregori et al., 2003; You et al., 2005). Effector T cells become increasingly resistant to nTreg suppression as the diabetogenic response progresses in NOD mice. The molecular basis for this resistant phenotype has yet to be elucidated; although effector T cells from older NOD mice exhibit a reduced sensitivity to the suppressive effects of TGFβ *in vitro* (You et al., 2005). Whether temporal changes in expression and/or signaling by TGFβ receptors account for the resistant phenotype needs to be determined. One interesting possibility is that the mild lymphopenia that has been reported in NOD mice, also affects the sensitivity of effector T cells to immunoregulation. In general, lymphopenia has been detected in patients with various autoimmune diseases including T1D, systemic lupus erythematosus, and rheumatoid arthritis (Marleau and Sarvetnick, 2005). It is believed that homeostatic expansion drives the development of increased numbers of pathogenic T effectors in the periphery. The mild lymphopenia in NOD mice is associated with the *idd3* genetic locus, and influenced by the elevated levels of IL-21 (King et al., 2004). Here, IL-21 is proposed to drive rapid T cell proliferation without survival, and therefore enhance T cell turnover to establish lymphopenia (King et al., 2004). Notably, T cells become resistant to certain tolerizing events under lymphopenic conditions. For instance, T cells expressing a β cell-specific transgenic TCR when adoptively transferred into NOD.*scid* recipients are less sensitive to antigen-induced tolerance (Long et al., 2006), and residual T cells following lymphodepletion in a cardiac allograft model are resistant to co-stimulatory blockade (Wu et al., 2004). Therefore one intriguing possibility is that T effector cells become resistant to nTreg-mediated immunoregulation under lymphopenic conditions. Of note is a recent study which showed that CD4$^+$ CD25$^-$ T cells also become resistant nTreg suppressor activity when conditioned with IL-21 (Clough et al., 2008).

4.2.1.2. Conventional Treg Unlike nTreg, "conventional" Treg (cTreg) differentiate from naïve CD4$^+$ T cells. Several subsets of cTreg exist that generally are defined by the types of cytokines expressed (Shevach, 2006). Upon antigen stimulation, cTreg subset differentiation is dependent on the cytokine *milieu* in which T cells are found, although other factors influence this process. cTreg include Th2-like cells characterized by IL-4 secretion; IL-4 is also critical for upregulation of the transcription factor GATA-3 and subsequent Th2 cell differentiation (Ansel et al., 2006). Th3-like cells, characterized by TGFβ secretion, are generally associated with antigen-stimulation at mucosal surfaces (Chen et al., 1994). A number of subsets of IL-10 secreting cTreg (IL10$^+$ cTreg) that vary in both the mode of induction and/or effector function have been defined (Roncarolo et al., 2006). For instance IL10$^+$ cTreg differentiation can be induced by IL-10 or via TGFβ and IL-27 in the absence of IL-10 (Awasthi et al., 2007; Barrat et al., 2002; Groux et al., 1997). Immunoregulation by IL10$^+$ cTreg is typically mediated by secretion of IL-10 which can inhibit IFNγ production by type 1 effector T cells, and activation and function of APC such as DC (Bhattacharyya et al., 2004; Moore et al., 2001). However, IL10$^+$ cTreg can also mediate a suppressor effect by cell-cell contact independent of IL-10 secretion (Vieira et al., 2004).

CD4$^+$ T cells prepared from peripheral blood of diabetic patients predominately secrete IFNγ in response to proinsulin, GAD65 and IA-2 peptides (Arif et al., 2004). In contrast, IL10$^+$ cTreg-like responses are detected in HLA-matched individuals suggesting a functional imbalance between Th1 and cTreg effectors in diabetics (Arif et al., 2004). Similarly, the reduced disease incidence in NOD male (~20%) versus female (~80%) mice corresponds with an increased frequency of IL-4 secreting Th2 cells residing in the islets of young animals (Fox and Danska, 1997). It is noteworthy that β cell-specific IL10$^+$ cTreg are found in NOD mice even at a late preclinical stage of T1D (You et al., 2004). This finding indicates that aberrant immunoregulation of β cell autoimmunity is not due to the absence of cTreg, but is attributed more so with an insufficient frequency of cTreg. The frequency of cTreg is partly dependent on the size of the pool of naïve precursors for β cell-specific clonotypes. This pool is depleted, however, as the diabetogenic response progresses and proinflammatory conditions favor the development of Th1 effectors. Consequently, the production of cTreg becomes progressively limited, and immunoregulation by cTreg less efficient.

The apparent skewing towards pathogenic Th1 effector versus cTreg differentiation for the most part appears to be independent of defects intrinsic to naïve T precursors. Although work by Koarada et al. suggests that such a defect may in fact exist in NOD CD4$^+$ T cells (Koarada et al., 2002). This group showed that naïve CD4$^+$ T cells prepared from NOD mice secrete increased IFNγ and reduced IL-4 and IL-10 when stimulated with anti-CD3 and -CD28 antibodies under "unbiased" culture conditions, compared to CD4$^+$ T cells isolated from IA^{g7}-expressing C57BL/6 mice. The basis for this preferential Th1 subset differentiation by NOD CD4$^+$ T cells still needs to be defined.

A number of factors extrinsic to T cells are believed to influence subset differentiation such as the nature of the TCR and pMHC interaction, and the magnitude of T cell co-stimulation. Diabetes is prevented in NOD mice expressing a IAβ^{g7} chain transgene in which the allele-defining histidine and aspartic acid residues found at positions 56 and 57, are substituted with proline and aspartic acid, respectively; protection is due to the induction of β cell-specific IL-4 and IL-10 secreting T effectors (Quartey-Papafio et al., 1995; Singer et al., 1998). Since the mutated IA^{g7PD} molecule binds a distinct profile of peptides compared to IA^{g7} (Suri et al., 2002, 2003), it is possible that changes in the selected TCR repertoire coupled with an altered repertoire of β cell-specific peptides promotes TCR signaling that favors cTreg differentiation (Tamura et al., 2004). Various studies have also demonstrated that differentiation of naïve CD4$^+$ T cells into Th2 versus Th1 subsets requires "stronger" CD28-mediated co-stimulation (Gudmundsdottir and Turka, 2001). In this regard it is noteworthy that treatment of neonatal NOD mice with an agonistic anti-CD28 antibody prevents diabetes by induction of Th2 cells, suggesting that CD28-mediated signaling is deficient in NOD mice (Arreaza et al., 1997).

The dominant factor responsible for promoting differentiation of pathogenic type 1 effectors in T1D is most likely the extracellular cytokine *milieu* at sites of T cell priming, such as the panLN and/or islets. The proinflammatory *milieu* in large part is governed by the activation and maturation status of resident APC such as DC. The role for DC in promoting pathogenic type 1 effectors will be discussed below in greater detail. Other cell types such as invariant NKT cells (iNKT cells) may also impact the cytokine environment of the panLN and islets. iNKT cells in mice are characterized by expression of a CD1d-restricted TCR consisting of an invariant Vα14-Jα1 chain and a limited number of TCR β chains (Kronenberg, 2005). Upon activation by glycolipids bound by CD1d, iNKT cells are potent producers of IL-4, IL-13 and IFNγ, and induce robust Th2-like reactivity *in vivo* (Kronenberg, 2005). The reduced numbers of iNKT cells in NOD mice (and possibly in diabetic patients) are believed to in part limit β cell-specific Th2 cell differentiation (Novak et al., 2007). Indeed treatment with the glycolipid α-galactosylceramide (αGalSer) expands iNKT cells in the panLN and protects NOD mice from diabetes (Hong et al., 2001; Sharif et al., 2001). In addition to directly conditioning the cytokine *milieu*, αGalSer-induced iNKT cells suppress β cell autoimmunity via multiple mechanisms including establishment of "tolergenic" DC (Naumov et al., 2001).

4.2.2. DC-mediated immunoregulation

DC are most often thought of as potent innate inducers of T cell-mediated immunity. However, DC also play an important role in the establishment and/or maintenance of peripheral T cell self-tolerance (Steinman, 2007; Steinman et al., 2005). Under homeostatic or noninflammatory conditions immature DC express low levels of CD40, CD80, CD86 and MHC class II, and readily endocytose antigen (Steinman, 2007). Upon activation by a variety of stimuli, DC maturation is induced reflected by up-regulation of co-stimulatory molecule and pMHC expression, and a concomitant increase in the capacity to stimulate naïve T cells (Steinman, 2007). Mature DC influence the extracellular *milieu* by secreting IL-12p70, TNFα, IFNγ, and IL-1 (Steinman, 2007). IL-12p70 promotes differentiation of type 1 $CD4^+$ and $CD8^+$ CTL effectors; notably TNFα, IFNγ, and IL-1 have direct cytotoxic effects on β cells *in vitro* (Cnop et al., 2005; Thomas and Kay, 2000; Wachlin et al., 2003). The immature DC phenotype under homeostatic conditions is "tolergenic" since T cells recognizing self-pMHC complexes fail to receive sufficient co-stimulation and undergo anergy and/or AICD (Hernandez et al., 2001). On the other hand, immature DC conditioned by IL-10, TGFβ, or apoptotic bodies are resistant to maturation-inducing stimuli, and actively maintain the tolergenic phenotype (Wallet et al., 2005). Maintenance of this tolergenic phenotype is in part due to a block in activation of NF-κB, a critical transcription factor which drives the expression of many genes involved in DC activation, maturation and effector function (Bhattacharyya et al., 2004; Sen et al., 2007).

Mature DC also contribute to tolerance by: (1) directly inducing apoptosis in T cells, (2) inducing/expanding Treg, in addition to (3) establishing a non-proinflammatory *milieu*. For instance, mature DC expressing PDL-1 induce apoptosis in T cells by binding PD-1 (Hochweller and Anderton, 2005). Furthermore, mature DC can expand nTreg in an antigen-specific manner in the presence of TGFβ and retinoic acid (Coombes et al., 2007). Moreover, mature DC under certain conditions secrete IL-10 or indoleamine 2,3-dioxygenase (IDO) (Mellor and Munn, 2004; Rutella et al., 2006). IL-10 blocks activation of neighboring APC and promotes differentiation of $IL-10^+$ aTreg (Rutella et al., 2006). IDO production by mature DC inhibits T cell proliferation through tryptophan catabolism (Mellor and Munn, 2004). Induction of IDO secretion by DC provides an interesting example of how nTreg or iTreg can modify DC effector function. nTreg or iTreg elicit IDO production by DC, by binding CD80/CD86 via CTLA-4 in the presence of IFNγ.

A number of observations indicate an important role for DC in initiating and maintaining β cell autoimmunity. Notably, DC are the first cells found infiltrating the islets of NOD mice (Nikolic et al., 2005). It is believed that physiological cell death within the pancreas, due to tissue remodeling at 12 days of age, triggers DC trafficking into the islets (Turley et al., 2003). A similar "ripple of death" for β cells occurs at birth in humans (Trudeau et al., 2000). Turley et al. demonstrated that these DC after entering the islets, endocytose β cell proteins and then traffick to the panLN to stimulate β cell-specific T cells in NOD mice (Turley et al., 2003). Saxena et al. have recently shown that selective ablation of DC via genetic means in NOD mice blocks activation of β cell-specific CD4$^+$ T cells and the development of insulitis (Saxena et al., 2007). Importantly, adoptive transfer of DC into NOD mice depleted of DC restores β cell-specific T cell reactivity (Saxena et al., 2007).

A variety of defects in DC have been identified in NOD mice and diabetic patients, which in turn may contribute to the preferential induction of a proinflammatory response. Various groups have shown that bone marrow-cultured and *ex vivo* DC prepared from NOD versus nonautoimmune mice exhibit a "hyperinflammatory" phenotype following various types of stimulation (Poligone et al., 2002; Weaver et al., 2001; Wheat et al., 2004). The latter is characterized by increased upregulation of CD40 and CD80, enhanced secretion of proinflammatory cytokines such as IL-12p70 and TNFα, and an increased capacity to stimulate CD4$^+$ and CD8$^+$T cells (Poligone et al., 2002; Weaver et al., 2001; Wheat et al., 2004). This phenotype is attributed to hyperactivation of the NF-κB pathway. Both the magnitude and persistence of NF-κB activation are enhanced in NOD DC compared to DC prepared from nonautoimmune mice (Poligone et al., 2002; Weaver et al., 2001; Wheat et al., 2004). It is noteworthy that macrophages from NOD mice also exhibit increased IL-12p70 secretion due to aberrant NF-κB activation (Liu and Beller, 2003; Sen et al., 2003), suggesting that dysregulation of this key transcription factor is a common defect among NOD APC. Elevated secretion of IL-12p70 and TNFα by NOD DC residing in the panLN and islets would be expected to drive type 1 subset differentiation, and enhance β cell destruction, respectively. In addition to promoting a proinflammatory *milieu*, NOD DC produce only low levels of IDO (Grohmann et al., 2003), which would further favor the development of pathogenic type 1 effector T cells.

Study of monocyte-derived DC from diabetic patients suggests a very different scenario than that observed in NOD mice. Various groups have reported that DC from diabetics exhibit mostly an immature phenotype and a decreased T cell stimulatory capacity compared to at risk or healthy individuals (Angelini et al., 2005; Mollah et al., 2008; Takahashi et al., 1998; Zacher et al., 2002). An immature phenotype could lead to a reduced capacity of DC to induce and/or expand different subsets of Treg. Interestingly, the relatively immature phenotype of DC from diabetic patients corresponds with defective activation of the NF-κB pathway following lipopolysaccharide stimulation (Mollah et al., 2008). Dysregulation of the pathway is in part due to over expression of src homology 2 domain-containing protein tyrosine phosphatase (SHP-1), a negative regulator of NF-κB (Mollah et al., 2008). This observation coupled with findings made in NOD mice indicate that dysregulation of the NF-κB pathway can occur in disparate ways that nevertheless influence the induction of self-tolerance.

5. SUMMARY

The development and expansion of β cell-specific pathogenic T effectors involve multiple events both intrinsic and extrinsic to T cells (Fig. 5.1). In the thymus, development of β cell-specific T precursors is believed to be due to resistance of thymocytes to induction of apoptosis, coupled with aberrant presentation of self-pMHC complexes. The latter is attributed to properties inherent to disease-linked HLA/MHC molecules that reduce the avidity of TCR and self-pMHC interactions, coupled with variations in TSA expression and presentation in the thymus. The consequence of these events is the production of an increased frequency of β cell-specific T precursors, which exit to the periphery.

In the periphery T cell intrinsic defects enhance the sensitivity of naïve or memory T cells to activating stimuli, in addition to promoting resistance to suppression by FoxP3-expressing Treg. Expansion of these pathogenic effectors is further permitted due to insufficient immunoregulation. Aberrant immunoregulation is marked by: (1) a progressive loss of nTreg function and impaired survival, and (2) limited production of cTreg. Dysregulation of maturation and/or function of DC, macrophages and other immune effector cells establishes an extracellular *milieu* that favors the differentiation and expansion of pathogenic effector T cells, and/or fails to effectively induce nTreg/iTreg and cTreg.

The relative contribution of the respective defects in central and peripheral tolerance in driving β cell autoimmunity remains unclear. Furthermore, as of yet defined forms of dysregulation surely contribute to the breakdown of β cell-specific tolerance. It is of particular interest that emerging clinical data suggests that multiple subtypes of T1D exist, which can be defined by age of onset, distinct genotypes and various environmental exposures. It is likely that these T1D subtypes will reflect distinct combinations of dysregulated events, which may influence central and/or peripheral tolerance to varying degrees qualitatively and/or quantitatively.

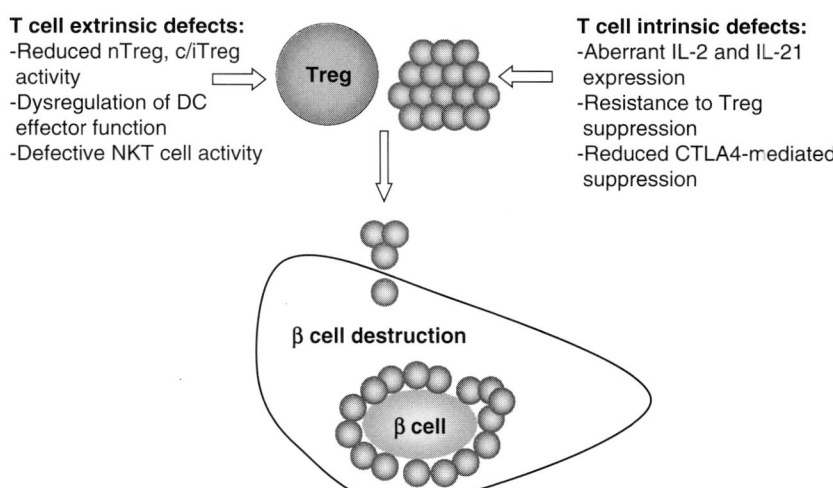

FIGURE 5.1 β cell autoimmunity results from dysregulation of central and peripheral T cell tolerance inducing events.

ACKNOWLEDGMENTS

This work was supported by a grant from the National Institutes of Health (R01AI05014). B. W. was supported by an ADA Career Development Award (1-04-CD-09).

REFERENCES

Alard, P., Manirarora, J. N., Parnell, S. A., Hudkins, J. L., Clark, S. L., and Kosiewicz, M. M. (2006). Deficiency in NOD antigen-presenting cell function may be responsible for suboptimal CD4 + CD25 + T-cell-mediated regulation and type 1 diabetes development in NOD mice. *Diabetes* **55**, 2098–2105.

Amrani, A., Verdaguer, J., Serra, P., Tafuro, S., Tan, R., and Santamaria, P. (2000). Progression of autoimmune diabetes driven by avidity maturation of a T-cell population. *Nature* **406**, 739–742.
Anderson, M. S., and Bluestone, J. A. (2005). The NOD mouse: A model of immune dysregulation. *Annu. Rev. Immunol.* **23**, 447–485.
Anderson, M. S., Venanzi, E. S., Klein, L., Chen, Z., Berzins, S. P., Turley, S. J., von Boehmer, H., Bronson, R., Dierich, A., Benoist, C., and Mathis, D. (2002). Projection of an immunological self shadow within the thymus by the aire protein. *Science* **298**, 1395–1401.
Andre, I., Gonzalez, A., Wang, B., Katz, J., Benoist, C., and Mathis, D. (1996). Checkpoints in the progression of autoimmune disease: Lessons from diabetes models. *Proc. Natl. Acad. Sci. USA* **93**, 2260–2263.
Angelini, F., Del Duca, E., Piccinini, S., Pacciani, V., Rossi, P., and Manca Bitti, M. L. (2005). Altered phenotype and function of dendritic cells in children with type 1 diabetes. *Clin. Exp. Immunol.* **142**, 341–346.
Ansari, M. J., Salama, A. D., Chitnis, T., Smith, R. N., Yagita, H., Akiba, H., Yamazaki, T., Azuma, M., Iwai, H., Khoury, S. J., Auchincloss, H., Jr., and Sayegh, M. H. (2003). The programmed death-1 (PD-1) pathway regulates autoimmune diabetes in nonobese diabetic (NOD) mice. *J. Exp. Med.* **198**, 63–69.
Ansel, K. M., Djuretic, I., Tanasa, B., and Rao, A. (2006) Regulation of Th2 differentiation and Il4 locus accessibility. *Annu. Rev. Immunol.* **24**, 607–656.
Arif, S., Tree, T. I., Astill, T. P., Tremble, J. M., Bishop. A. J., Dayan, C. M., Roep, B. O., and Peakman, M. (2004). Autoreactive T cell responses show proinflammatory polarization in diabetes but a regulatory phenotype in health. *J. Clin. Invest.* **113**, 451–463.
Arreaza, G. A., Cameron, M. J., Jaramillo, A., Gill, B. M., Hardy, D., Laupland, K. B., Rapoport, M. J., Zucker, P., Chakrabarti, S., Chensue, S. W., Qin, H. Y., Singh, B., *et al.* (1997). Neonatal activation of CD28 signaling overcomes T cell anergy and prevents autoimmune diabetes by an IL-4-dependent mechanism. *J. Clin. Invest.* **100**, 2243–2253.
Arreaza, G., Salojin, K., Yang, W., Zhang, J., Gill, B., Mi, Q. S., Gao, J. X., Meagher, C., Cameron, M., and Delovitch, T. L. (2003). Deficient activation and resistance to activation-induced apoptosis of CD8 + T cells is associated with defective peripheral tolerance in nonobese diabetic mice. *Clin. Immunol.* **107**, 103–115.
Awasthi, A., Carrier, Y., Peron, J. P., Bettelli, E., Kamanaka, M., Flavell, R. A., Kuchroo, V. K., Oukka, M., and Weiner, H. L. (2007). A dominant function for interleukin 27 in generating interleukin 10-producing anti-inflammatory T cells. *Nat. Immunol.* **8**, 1380–1389.
Bach, J. F. (1994). Insulin-dependent diabetes mellitus as an autoimmune disease. *Endocr. Rev.* **15**, 516–542.
Barrat, F. J., Cua, D. J., Boonstra, A., Richards, D. F., Crain, C., Savelkoul, H. F., de Waal-Malefyt, R., Coffman, R. L., Hawrylowicz, C. M., and O'Garra, A. (2002). *In vitro* generation of interleukin 10-producing regulatory CD4(+) T cells is induced by immunosuppressive drugs and inhibited by T helper type 1 (Th1)- and Th2-inducing cytokines. *J. Exp. Med.* **195**, 603–616.
Bhattacharyya, S., Sen, P., Wallet, M., Long, B., Baldwin, A. S., Jr., and Tisch, R. (2004). Immunoregulation of dendritic cells by IL-10 is mediated through suppression of the PI3K/Akt pathway and of IkappaB kinase activity. *Blood* **104**, 1100–1109.
Bottini, N., Vang, T., Cucca, F., and Mustelin, T. (2006). Role of PTPN22 in type 1 diabetes and other autoimmune diseases. *Semin. Immunol.* **18**, 207–213.
Brusko, T. M., Wasserfall, C. H., Clare-Salzler, M. J., Schatz, D. A., and Atkinson, M. A. (2005). Functional defects and the influence of age on the frequency of CD4 + CD25 + T-cells in type 1 diabetes. *Diabetes* **54**, 1407–1414.
Chen, L. (2004). Co-inhibitory molecules of the B7-CD28 family in the control of T-cell immunity. *Nat. Rev. Immunol.* **4**, 336–347.

Chen, Y., Kuchroo, V. K., Inobe, J., Hafler, D. A., and Weiner, H. L. (1994). Regulatory T cell clones induced by oral tolerance: Suppression of autoimmune encephalomyelitis. *Science* **265**, 1237–1240.

Chen, W., Jin, W., Hardegen, N., Lei, K. J., Li, L., Marinos, N., McGrady, G., and Wahl, S. M. (2003). Conversion of peripheral CD4 + CD25- naive T cells to CD4 + CD25 + regulatory T cells by TGF-beta induction of transcription factor Foxp3. *J. Exp. Med.* **198**, 1875–1886.

Chen, Z., Herman, A. E., Matos, M., Mathis, D., and Benoist, C. (2005). Where CD4 + CD25 + T reg cells impinge on autoimmune diabetes. *J. Exp. Med.* **202**, 1387–1397.

Chentoufi, A. A., and Polychronakos, C. (2002). Insulin expression levels in the thymus modulate insulin-specific autoreactive T-cell tolerance: The mechanism by which the IDDM2 locus may predispose to diabetes. *Diabetes* **51**, 1383–1390.

Clough, L. E., Wang, C. J., Schmidt, E. M., Booth, G., Hou, T. Z., Ryan, G. A., and Walker, L. S. (2008). Release from regulatory T cell-mediated suppression during the onset of tissue-specific autoimmunity is associated with elevated IL-21. *J. Immunol.* **180**, 5393–5401.

Cnop, M., Welsh, N., Jonas, J. C., Jorns, A., Lenzen, S., and Eizirik, D. L. (2005). Mechanisms of pancreatic beta-cell death in type 1 and type 2 diabetes: Many differences, few similarities. *Diabetes* **54**(Suppl. 2), S97–S107.

Coombes, J. L., Siddiqui, K. R., Arancibia-Carcamo, C. V., Hall, J., Sun, C. M., Belkaid, Y., and Powrie, F. (2007). A functionally specialized population of mucosal CD103 + DCs induces Foxp3 + regulatory T cells via a TGF-beta and retinoic acid-dependent mechanism. *J. Exp. Med.* **204**, 1757–1764.

Davidson, T. S., DiPaolo, R. J., Andersson, J., and Shevach, E. M. (2007). Cutting Edge: IL-2 is essential for TGF-beta-mediated induction of Foxp3 + T regulatory cells. *J. Immunol.* **178**, 4022–4026.

Davis, S. J., and van der Merwe, P. A. (2006). The kinetic-segregation model: TCR triggering and beyond. *Nat. Immunol.* **7**, 803–809.

Decallonne, B., van Etten, E., Giulietti, A., Casteels, K., Overbergh, L., Bouillon, R., and Mathieu, C. (2003). Defect in activation-induced cell death in non-obese diabetic (NOD) T lymphocytes. *J. Autoimmun.* **20**, 219–226.

Derbinski, J., Gabler, J., Brors, B., Tierling, S., Jonnakuty, S., Hergenhahn, M., Peltonen, L., Walter, J., and Kyewski, B. (2005). Promiscuous gene expression in thymic epithelial cells is regulated at multiple levels. *J. Exp. Med.* **202**, 33–45.

Eisenbarth, G. S. (2004). Prediction of type 1 diabetes: The natural history of the prediabetic period. *Adv. Exp. Med. Biol.* **552**, 268–290.

Fehervari, Z., Yamaguchi, T., and Sakaguchi, S. (2006). The dichotomous role of IL-2: Tolerance versus immunity. *Trends Immunol.* **27**, 109–111.

Fontenot, J. D., Gavin, M. A., and Rudensky, A. Y. (2003). Foxp3 programs the development and function of CD4 + CD25 + regulatory T cells. *Nat. Immunol.* **4**, 330–336.

Fox, C. J., and Danska, J. S. (1997). IL-4 expression at the onset of islet inflammation predicts nondestructive insulitis in nonobese diabetic mice. *J. Immunol.* **158**, 2414–2424.

Gambineri, E., Torgerson, T. R., and Ochs, H. D. (2003). Immune dysregulation, polyendocrinopathy, enteropathy, and X-linked inheritance (IPEX), a syndrome of systemic autoimmunity caused by mutations of FOXP3, a critical regulator of T-cell homeostasis. *Curr. Opin. Rheumatol.* **15**, 430–435.

Gregg, R. K., Jain, R., Schoenleber, S. J., Divekar, R., Bell, J. J., Lee, H. H., Yu, P., and Zaghouani, H. (2004). A sudden decline in active membrane-bound TGF-beta impairs both T regulatory cell function and protection against autoimmune diabetes. *J. Immunol.* **173**, 7308–7316.

Gregori, S., Giarratana, N., Smiroldo, S., and Adorini, L. (2003). Dynamics of pathogenic and suppressor T cells in autoimmune diabetes development. *J. Immunol.* **171**, 4040–4047.

Grohmann, U., Fallarino, F., Bianchi, R., Orabona, C., Vacca, C., Fioretti, M. C., and Puccetti, P. (2003). A defect in tryptophan catabolism impairs tolerance in nonobese diabetic mice. *J. Exp. Med.* **198**, 153–160.

Groux, H., O'Garra, A., Bigler, M., Rouleau, M., Antonenko, S., de Vries, J. E., and Roncarolo, M. G. (1997). A CD4 + T-cell subset inhibits antigen-specific T-cell responses and prevents colitis. *Nature* **389**, 737–742.
Gudmundsdottir, H., and Turka, L. A. (2001). A closer look at homeostatic proliferation of CD4 + T cells: Costimulatory requirements and role in memory formation. *J. Immunol.* **167**, 3699–3707.
Herman, A. E., Freeman, G. J., Mathis, D., and Benoist, C. (2004). CD4 + CD25 + T regulatory cells dependent on ICOS promote regulation of effector cells in the prediabetic lesion. *J. Exp. Med.* **199**, 1479–1489.
Hernandez, J., Aung, S., Redmond, W. L., and Sherman, L. A. (2001). Phenotypic and functional analysis of CD8(+) T cells undergoing peripheral deletion in response to cross-presentation of self-antigen. *J. Exp. Med.* **194**, 707–717.
Hill, J. A., Feuerer, M., Tash, K., Haxhinasto, S., Perez, J., Melamed, R., Mathis, D., and Benoist, C. (2007). Foxp3 transcription-factor-dependent and -independent regulation of the regulatory T cell transcriptional signature. *Immunity* **27**, 786–800.
Hochweller, K., and Anderton, S. M. (2005). Kinetics of costimulatory molecule expression by T cells and dendritic cells during the induction of tolerance versus immunity *in vivo*. *Eur. J. Immunol.* **35**, 1086–1096.
Hoglund, P., Mintern, J., Waltzinger, C., Heath, W., Benoist, C., and Mathis, D. (1999). Initiation of autoimmune diabetes by developmentally regulated presentation of islet cell antigens in the pancreatic lymph nodes. *J. Exp. Med.* **189**, 331–339.
Hong, S., Wilson, M. T., Serizawa, I., Wu, L., Singh, N., Naidenko, O. V., Miura, T., Haba, T., Scherer, D. C., Wei, J., Kronenberg, M., Koezuka, Y., et al. (2001). The natural killer T-cell ligand alpha-galactosylceramide prevents autoimmune diabetes in non-obese diabetic mice. *Nat. Med.* **7**, 1052–1056.
Hori, S., Nomura, T., and Sakaguchi, S. (2003). Control of regulatory T cell development by the transcription factor Foxp3. *Science* **299**, 1057–1061.
Kang, S. M., Tang, Q., and Bluestone, J. A. (2007). CD4 + CD25 + regulatory T cells in transplantation: Progress, challenges and prospects. *Am. J. Transplant.* **7**, 1457–1463.
Kaufman, D. L., Clare-Salzler, M., Tian, J., Forsthuber, T., Ting, G. S., Robinson, P., Atkinson, M. A., Sercarz, E. E., Tobin, A. J., and Lehmann, P. V. (1993). Spontaneous loss of T-cell tolerance to glutamic acid decarboxylase in murine insulin-dependent diabetes. *Nature* **366**, 69–72.
Keir, M. E., Liang, S. C., Guleria, I., Latchman, Y. E., Qipo, A., Albacker, L. A., Koulmanda, M., Freeman, G. J., Sayegh, M. H., and Sharpe, A. H. (2006). Tissue expression of PD-L1 mediates peripheral T cell tolerance. *J. Exp. Med.* **203**, 883–895.
Khattri, R., Cox, T., Yasayko, S. A., and Ramsdell, F. (2003). An essential role for Scurfin in CD4 + CD25 + T regulatory cells. *Nat. Immunol.* **4**, 337–342.
Kim, J. M., Rasmussen, J. P., and Rudensky, A. Y. (2007). Regulatory T cells prevent catastrophic autoimmunity throughout the lifespan of mice. *Nat. Immunol.* **8**, 191–197.
King, C., Ilic, A., Koelsch, K., and Sarvetnick, N. (2004). Homeostatic expansion of T cells during immune insufficiency generates autoimmunity. *Cell* **117**, 265–277.
Koarada, S., Wu, Y., Olshansky, G., and Ridgway, W. M. (2002). Increased nonobese diabetic Th1:Th2 (IFN-gamma:IL-4) ratio is CD4 + T cell intrinsic and independent of APC genetic background. *J. Immunol.* **169**, 6580–6587.
Korman, A. J., Peggs, K. S., and Allison, J. P. (2006). Checkpoint blockade in cancer immunotherapy. *Adv. Immunol.* **90**, 297–339.
Korn, T., Bettelli, E., Gao, W., Awasthi, A., Jager, A., Strom, T. B., Oukka, M., and Kuchroo, V. K. (2007). IL-21 initiates an alternative pathway to induce proinflammatory T(H)17 cells. *Nature* **448**, 484–487.
Krishnamurthy, B., Dudek, N. L., McKenzie, M. D., Purcell, A. W., Brooks, A. G., Gellert, S., Colman, P. G., Harrison, L. C., Lew, A. M., Thomas, H. E., and Kay, T. W. (2006).

Responses against islet antigens in NOD mice are prevented by tolerance to proinsulin but not IGRP. *J. Clin. Invest.* **116,** 3258–3265.

Kristiansen, O. P., Larsen, Z. M., and Pociot, F. (2000). CTLA-4 in autoimmune diseases—a general susceptibility gene to autoimmunity? *Genes Immun.* **1,** 170–184.

Kronenberg, M. (2005). Toward an understanding of NKT cell biology: Progress and paradoxes. *Annu. Rev. Immunol.* **23,** 877–900.

Krummel, M. F., and Allison, J. P. (1995). CD28 and CTLA-4 have opposing effects on the response of T cells to stimulation. *J. Exp. Med.* **182,** 459–465.

Li, Y., and Yee, C. (2008). IL-21 mediated Foxp3 suppression leads to enhanced generation of antigen-specific CD8 + cytotoxic T lymphocytes. *Blood* **111,** 229–235.

Liblau, R. S., Singer, S. M., and McDevitt, H. O. (1995). Th1 and Th2 CD4 + T cells in the pathogenesis of organ-specific autoimmune diseases. *Immunol. Today* **16,** 34–38.

Lindley, S., Dayan, C. M., Bishop, A., Roep, B. O., Peakman, M., and Tree, T. I. (2005). Defective suppressor function in CD4(+)CD25(+) T-cells from patients with type 1 diabetes. *Diabetes* **54,** 92–99.

Liston, A., Lesage, S., Wilson, J., Peltonen, L., and Goodnow, C. C. (2003). Aire regulates negative selection of organ-specific T cells. *Nat. Immunol.* **4,** 350–354.

Liston, A., Lesage, S., Gray, D. H., O'Reilly, L. A., Strasser, A., Fahrer, A. M., Boyd, R. L., Wilson, J., Baxter, A. G., Gallo, E. M., Crabtree, G. R., Peng, K., *et al.* (2004). Generalized resistance to thymic deletion in the NOD mouse; a polygenic trait characterized by defective induction of Bim. *Immunity* **21,** 817–830.

Liu, J., and Beller, D. I. (2003). Distinct pathways for NF-kappa B regulation are associated with aberrant macrophage IL-12 production in lupus- and diabetes-prone mouse strains. *J. Immunol.* **170,** 4489–4496.

Long, B., Wong, C. P., Wang, Y., and Tisch, R. (2006). Lymphopenia-driven CD8(+) T cells are resistant to antigen-induced tolerance in NOD.scid mice. *Eur. J. Immunol.* **36,** 2003–2012.

Lowe, C. E., Cooper, J. D., Brusko, T., Walker, N. M., Smyth, D. J., Bailey, R., Bourget, K., Plagnol, V., Field, S., Atkinson, M., Clayton, D. G., and Wicker, L. S. *et al.* (2007). Large-scale genetic fine mapping and genotype-phenotype associations implicate polymorphism in the IL2RA region in type 1 diabetes. *Nat. Genet.* **39,** 1074–1082.

Luhder, F., Hoglund, P., Allison, J. P., Benoist, C., and Mathis, D. (1998). Cytotoxic T lymphocyte-associated antigen 4 (CTLA-4) regulates the unfolding of autoimmune diabetes. *J. Exp. Med.* **187,** 427–432.

Magistrelli, G., Jeannin, P., Herbault, N., Benoit De Coignac, A., Gauchat, J. F., Bonnefoy, J. Y., and Delneste, Y. (1999). A soluble form of CTLA-4 generated by alternative splicing is expressed by nonstimulated human T cells. *Eur. J. Immunol.* **29,** 3596–3602.

Maier, L. M., and Wicker, L. S. (2005). Genetic susceptibility to type 1 diabetes. *Curr. Opin. Immunol.* **17,** 601–608.

Marleau, A. M., and Sarvetnick, N. (2005). T cell homeostasis in tolerance and immunity. *J. Leukoc. Biol.* **78,** 575–584.

Mathis, D., and Benoist, C. (2007). A decade of AIRE. *Nat. Rev. Immunol.* **7,** 645–650.

Mellanby, R. J., Thomas, D., Phillips, J. M., and Cooke, A. (2007). Diabetes in non-obese diabetic mice is not associated with quantitative changes in CD4 + CD25 + Foxp3 + regulatory T cells. *Immunology* **121,** 15–28.

Mellor, A. L., and Munn, D. H. (2004). IDO expression by dendritic cells: Tolerance and tryptophan catabolism. *Nat. Rev. Immunol.* **4,** 762–774.

Mollah, Z. U., Pai, S., Moore, C., O'Sullivan, B. J., Harrison, M. J., Peng, J., Phillips, K., Prins, J. B., Cardinal, J., and Thomas, R. (2008). Abnormal NF-κB function characterizes human type 1 diabetes dendritic cells and monocytes. *J. Immunol.* **180,** 3166–3175.

Moore, K. W., de Waal Malefyt, R., Coffman, R. L., and O'Garra, A. (2001). Interleukin-10 and the interleukin-10 receptor. *Annu. Rev. Immunol.* **19,** 683–765.

Mueller, D. L. (2004). E3 ubiquitin ligases as T cell anergy factors. *Nat. Immunol.* **5,** 883–890.
Murakami, M., Sakamoto, A., Bender, J., Kappler, J., and Marrack, P. (2002). CD25 + CD4 + T cells contribute to the control of memory CD8 + T cells. *Proc. Natl. Acad. Sci. USA* **99,** 8832–8837.
Nakayama, M., Abiru, N., Moriyama, H., Babaya, N., Liu, E., Miao, D., Yu, L., Wegmann, D. R., Hutton, J. C., Elliott, J. F., and Eisenbarth, G. S. (2005). Prime role for an insulin epitope in the development of type 1 diabetes in NOD mice. *Nature* **435,** 220–223.
Nakayama, M., Beilke, J. N., Jasinski, J. M., Kobayashi, M., Miao, D., Li, M., Coulombe, M. G., Liu, E., Elliott, J. F., Gill, R. G., and Eisenbarth, G. S. (2007). Priming and effector dependence on insulin B:9–23 peptide in NOD islet autoimmunity. *J. Clin. Invest.* **117,** 1835–1843.
Naumov, Y. N., Bahjat, K. S., Gausling, R., Abraham, R., Exley, M. A., Koezuka, Y., Balk, S. B., Strominger, J. L., Clare-Salzer, M., and Wilson, S. B. (2001). Activation of CD1d-restricted T cells protects NOD mice from developing diabetes by regulating dendritic cell subsets. *Proc. Natl. Acad. Sci. USA* **98,** 13838–13843.
Nikolic, T., Geutskens, S. B., van Rooijen, N., Drexhage, H. A., and Leenen, P. J. (2005). Dendritic cells and macrophages are essential for the retention of lymphocytes in (peri)-insulitis of the nonobese diabetic mouse: A phagocyte depletion study. *Lab. Invest.* **85,** 487–501.
Noel, P. J., Boise, L. H., and Thompson, C. B. (1996). Regulation of T cell activation by CD28 and CTLA4. *Adv. Exp. Med. Biol.* **406,** 209–217.
Novak, J., Griseri, T., Beaudoin, L., and Lehuen, A. (2007). Regulation of type 1 diabetes by NKT cells. *Int. Rev. Immunol.* **26,** 49–72.
Ohashi, P. S., Oehen, S., Buerki, K., Pircher, H., Ohashi, C. T., Odermatt, B., Malissen, B., Zinkernagel, R. M., and Hengartner, H. (1991). Ablation of "tolerance" and induction of diabetes by virus infection in viral antigen transgenic mice. *Cell* **65,** 305–317.
Okazaki, T., and Honjo, T. (2007). PD-1 and PD-1 ligands: From discovery to clinical application. *Int. Immunol.* **19,** 813–824.
Onengut-Gumuscu, S., and Concannon, P. (2006). Recent advances in the immunogenetics of human type 1 diabetes. *Curr. Opin. Immunol.* **18,** 634–638.
Parry, R. V., Chemnitz, J. M., Frauwirth, K. A., Lanfranco, A. R., Braunstein, I., Kobayashi, S. V., Linsley, P. S., Thompson, C. B., and Riley, J. L. (2005). CTLA-4 and PD-1 receptors inhibit T-cell activation by distinct mechanisms. *Mol. Cell. Biol.* **25,** 9543–9553.
Piao, W. H., Jee, Y. H., Liu, R. L., Coons, S. W., Kala, M., Collins, M., Young, D. A., Campagnolo, D. I., Vollmer, T. L., Bai, X. F., *et al.* (2008). IL-21 modulates CD4 + CD25 + regulatory T-cell homeostasis in experimental autoimmune encephalomyelitis. *Scand. J. Immunol.* **67,** 37–46.
Poligone, B., Weaver, D. J., Jr., Sen, P., Baldwin, A. S., Jr., and Tisch, R. (2002). Elevated NF-kappaB activation in nonobese diabetic mouse dendritic cells results in enhanced APC function. *J. Immunol.* **168,** 188–196.
Pop, S. M., Wong, C. P., Culton, D. A., Clarke, S. H., and Tisch, R. (2005). Single cell analysis shows decreasing FoxP3 and TGFbeta1 coexpressing CD4 + CD25 + regulatory T cells during autoimmune diabetes. *J. Exp. Med.* **201,** 1333–1346.
Pugliese, A., and Diez, J. (2002). Lymphoid organs contain diverse cells expressing self-molecules. *Nat. Immunol.* **3,** 335–336; author reply.
Pugliese, A., Brown, D., Garza, D., Murchison, D., Zeller, M., Redondo, M. J., Diez, J., Eisenbarth, G. S., Patel, D. D., and Ricordi, C. (2001). Self-antigen-presenting cells expressing diabetes-associated autoantigens exist in both thymus and peripheral lymphoid organs. *J. Clin. Invest.* **107,** 555–564.

Quartey-Papafio, R., Lund, T., Chandler, P., Picard, J., Ozegbe, P., Day, S., Hutchings, P. R., O'Reilly, L., Kioussis, D., Simpson, E., et al. (1995). Aspartate at position 57 of nonobese diabetic I-Ag7 beta-chain diminishes the spontaneous incidence of insulin-dependent diabetes mellitus. *J. Immunol.* **154**, 5567–5575.

Roncarolo, M. G., Gregori, S., Battaglia, M., Bacchetta, R., Fleischhauer, K., and Levings, M. K. (2006). Interleukin-10-secreting type 1 regulatory T cells in rodents and humans. *Immunol. Rev.* **212**, 28–50.

Rutella, S., Danese, S., and Leone, G. (2006). Tolerogenic dendritic cells: Cytokine modulation comes of age. *Blood* **108**, 1435–1440.

Sakaguchi, S. (2005). Naturally arising Foxp3-expressing CD25 + CD4 + regulatory T cells in immunological tolerance to self and non-self. *Nat. Immunol.* **6**, 345–352.

Sarvetnick, N. (2000). Etiology of autoimmunity. *Immunol. Res.* **21**, 357–362.

Saverino, D., Brizzolara, R., Simone, R., Chiappori, A., Milintenda-Floriani, F., Pesce, G., and Bagnasco, M. (2007). Soluble CTLA-4 in autoimmune thyroid diseases: Relationship with clinical status and possible role in the immune response dysregulation. *Clin. Immunol.* **123**, 190–198.

Saxena, V., Ondr, J. K., Magnusen, A. F., Munn, D. H., and Katz, J. D. (2007). The countervailing actions of myeloid and plasmacytoid dendritic cells control autoimmune diabetes in the nonobese diabetic mouse. *J. Immunol.* **179**, 5041–5053.

Sen, P., Bhattacharyya, S., Wallet, M., Wong, C. P., Poligone, B., Sen, M., Baldwin, A. S., Jr., and Tisch, R. (2003). NF-kappa B hyperactivation has differential effects on the APC function of nonobese diabetic mouse macrophages. *J. Immunol.* **170**, 1770–1780.

Sen, P., Wallet, M. A., Yi, Z., Huang, Y., Henderson, M., Mathews, C. E., Earp, H. S., Matsushima, G., Baldwin, A. S., Jr., and Tisch, R. M. (2007). Apoptotic cells induce Mer tyrosine kinase-dependent blockade of NF-kappaB activation in dendritic cells. *Blood* **109**, 653–660.

Setoguchi, R., Hori, S., Takahashi, T., and Sakaguchi, S. (2005). Homeostatic maintenance of natural Foxp3(+) CD25(+) CD4(+) regulatory T cells by interleukin (IL)-2 and induction of autoimmune disease by IL-2 neutralization. *J. Exp. Med.* **201**, 723–735.

Sharif, S., Arreaza, G. A., Zucker, P., Mi, Q. S., Sondhi, J., Naidenko, O. V., Kronenberg, M., Koezuka, Y., Delovitch, T. L., Gombert, J. M., Leite-De-Moraes, M., Gouarin, C., et al. (2001). Activation of natural killer T cells by alpha-galactosylceramide treatment prevents the onset and recurrence of autoimmune Type 1 diabetes. *Nat. Med.* **7**, 1057–1062.

Shevach, E. M. (2006). From vanilla to 28 flavors: Multiple varieties of T regulatory cells. *Immunity* **25**, 195–201.

Singer, S. M., Tisch, R., Yang, X. D., Sytwu, H. K., Liblau, R., and McDevitt, H. O. (1998). Prevention of diabetes in NOD mice by a mutated I-Ab transgene. *Diabetes* **47**, 1570–1577.

Spolski, R., and Leonard, W. J. (2008). Interleukin-21: Basic biology and implications for cancer and autoimmunity. *Annu. Rev. Immunol.* **26**, 57–79.

Starr, T. K., Jameson, S. C., and Hogquist, K. A. (2003). Positive and negative selection of T cells. *Annu. Rev. Immunol.* **21**, 139–176.

Steinman, R. M. (2007). Dendritic cells: Understanding immunogenicity. *Eur. J. Immunol.* **37** (Suppl. 1), S53–S60.

Steinman, R. M., Bonifaz, L., Fujii, S., Liu, K., Bonnyay, D., Yamazaki, S., Pack, M., Hawiger, D., Iyoda, T., Inaba, K., and Nussenzweig, M. C. (2005). The innate functions of dendritic cells in peripheral lymphoid tissues. *Adv. Exp. Med. Biol.* **560**, 83–97.

Suri, A., Vidavsky, I., van der Drift, K., Kanagawa, O., Gross, M. L., and Unanue, E. R. (2002). In APCs, the autologous peptides selected by the diabetogenic I-Ag7 molecule are unique and determined by the amino acid changes in the P9 pocket. *J. Immunol.* **168**, 1235–1243.

Suri, A., Walters, J. J., Kanagawa, O., Gross, M. L., and Unanue, E. R. (2003). Specificity of peptide selection by antigen-presenting cells homozygous or heterozygous for expression of class II MHC molecules: The lack of competition. *Proc. Natl. Acad. Sci. USA* **100**, 5330–5335.

Suri, A., Levisetti, M. G., and Unanue, E. R. (2008). Do the peptide-binding properties of diabetogenic class II molecules explain autoreactivity? *Curr. Opin. Immunol.* **20**, 105–110.

Takahashi, K., Honeyman, M. C., and Harrison, L. C. (1998). Impaired yield, phenotype, and function of monocyte-derived dendritic cells in humans at risk for insulin-dependent diabetes. *J. Immunol.* **161**, 2629–2635.

Tamura, T., Ariga, H., Kinashi, T., Uehara, S., Kikuchi, T., Nakada, M., Tokunaga, T., Xu, W., Kariyone, A., Saito, T., Kitamura, T., Maxwell, G., et al. (2004). The role of antigenic peptide in CD4 + T helper phenotype development in a T cell receptor transgenic model. *Int. Immunol.* **16**, 1691–1699.

Tang, Q., and Bluestone, J. A. (2008). The Foxp3 + regulatory T cell: A jack of all trades, master of regulation. *Nat. Immunol.* **9**, 239–244.

Tang, Q., Adams, J. Y., Penaranda, C., Melli, K., Piaggio, E., Sgouroudis, E., Piccirillo, C. A., Salomon, B. L., and Bluestone, J. A. (2008). Central role of defective interleukin-2 production in the triggering of islet autoimmune destruction. *Immunity* **28**, 687–697.

Thomas, H. E., and Kay, T. W. (2000). Beta cell destruction in the development of autoimmune diabetes in the non-obese diabetic (NOD) mouse. *Diabetes Metab. Res. Rev.* **16**, 251–261.

Tian, J., Zekzer, D., Lu, Y., Dang, H., and Kaufman, D. L. (2006). B cells are crucial for determinant spreading of T cell autoimmunity among beta cell antigens in diabetes-prone nonobese diabetic mice. *J. Immunol.* **176**, 2654–2661.

Tisch, R., and McDevitt, H. (1996). Insulin-dependent diabetes mellitus. *Cell* **85**, 291–297.

Tisch, R., Yang, X. D., Singer, S. M., Liblau, R. S., Fugger, L., and McDevitt, H. O. (1993). Immune response to glutamic acid decarboxylase correlates with insulitis in non-obese diabetic mice. *Nature* **366**, 72–75.

Tivol, E. A., Borriello, F., Schweitzer, A. N., Lynch, W. P., Bluestone, J. A., and Sharpe, A. H. (1995). Loss of CTLA-4 leads to massive lymphoproliferation and fatal multiorgan tissue destruction, revealing a critical negative regulatory role of CTLA-4. *Immunity* **3**, 541–547.

Todd, J. A., and Wicker, L. S. (2001). Genetic protection from the inflammatory disease type 1 diabetes in humans and animal models. *Immunity* **15**, 387–395.

Tritt, M., Sgouroudis, E., d'Hennezel, E., Albanese, A., and Piccirillo, C. A. (2008). Functional waning of naturally occurring CD4 + regulatory T-cells contributes to the onset of autoimmune diabetes. *Diabetes* **57**, 113–123.

Trudeau, J. D., Dutz, J. P., Arany, E., Hill, D. J., Fieldus, W. E., and Finegood, D. T. (2000). Neonatal beta-cell apoptosis: A trigger for autoimmune diabetes? *Diabetes* **49**, 1–7.

Turley, S., Poirot, L., Hattori, M., Benoist, C., and Mathis, D. (2003). Physiological beta cell death triggers priming of self-reactive T cells by dendritic cells in a type-1 diabetes model. *J. Exp. Med.* **198**, 1527–1537.

Ueda, H., Howson, J. M., Esposito, L., Heward, J., Snook, H., Chamberlain, G., Rainbow, D. B., Hunter, K. M., Smith, A. N., Di Genova, G., Herr, M. H., Dahlman, I., et al. (2003). Association of the T-cell regulatory gene CTLA4 with susceptibility to autoimmune disease. *Nature* **423**, 506–511.

Venanzi, E. S., Benoist, C., and Mathis, D. (2004). Good riddance: Thymocyte clonal deletion prevents autoimmunity. *Curr. Opin. Immunol.* **16**, 197–202.

Vieira, P. L., Christensen, J. R., Minaee, S., O'Neill, E. J., Barrat, F. J., Boonstra, A., Barthlott, T., Stockinger, B., Wraith, D. C., and O'Garra, A. (2004). IL-10-secreting regulatory T cells do not express Foxp3 but have comparable regulatory function to naturally occurring CD4 + CD25 + regulatory T cells. *J. Immunol.* **172**, 5986–5993.

Vijayakrishnan, L., Slavik, J. M., Illes, Z., Greenwald, R. J., Rainbow, D., Greve, B., Peterson, L. B., Hafler, D. A., Freeman, G. J., Sharpe, A. H., Wicker, L. S., and Kuchroo, V. K. (2004). An autoimmune disease-associated CTLA-4 splice variant lacking the B7 binding domain signals negatively in T cells. *Immunity* **20**, 563–575.

von Herrath, M. G., Dockter, J., and Oldstone, M. B. (1994). How virus induces a rapid or slow onset insulin-dependent diabetes mellitus in a transgenic model. *Immunity* **1,** 231–242.
Wachlin, G., Augstein, P., Schroder, D., Kuttler, B., Kloting, I., Heinke, P., and Schmidt, S. (2003). IL-1beta, IFN-gamma and TNF-alpha increase vulnerability of pancreatic beta cells to autoimmune destruction. *J. Autoimmun.* **20,** 303–312.
Wallet, M. A., Sen, P., and Tisch, R. (2005). Immunoregulation of dendritic cells. *Clin. Med. Res.* **3,** 166–175.
Walunas, T. L., Bakker, C. Y., and Bluestone, J. A. (1996). CTLA-4 ligation blocks CD28-dependent T cell activation. *J. Exp. Med.* **183,** 2541–2550.
Weaver, D. J., Jr., Poligone, B., Bui, T., Abdel-Motal, U. M., Baldwin, A. S., Jr., and Tisch, R. (2001). Dendritic cells from nonobese diabetic mice exhibit a defect in NF-kappa B regulation due to a hyperactive I kappa B kinase. *J. Immunol.* **167,** 1461–1468.
Wheat, W., Kupfer, R., Gutches, D. G., Rayat, G. R., Beilke, J., Scheinman, R. I., and Wegmann, D. R. (2004). Increased NF-kappa B activity in B cells and bone marrow-derived dendritic cells from NOD mice. *Eur. J. Immunol.* **34,** 1395–1404.
Wicker, L. S., Todd, J. A., and Peterson, L. B. (1995). Genetic control of autoimmune diabetes in the NOD mouse. *Annu. Rev. Immunol.* **13,** 179–200.
Wicker, L. S., Clark, J., Fraser, H. I., Garner, V. E., Gonzalez-Munoz, A., Healy, B., Howlett, S., Hunter, K., Rainbow, D., Rosa, R. L., Smink, L. J., Todd, J. A., *et al.* (2005). Type 1 diabetes genes and pathways shared by humans and NOD mice. *J. Autoimmun.* **25**(Suppl), 29–33.
Wu, A. J., Hua, H., Munson, S. H., and McDevitt, H. O. (2002). Tumor necrosis factor-alpha regulation of CD4 + CD25 + T cell levels in NOD mice. *Proc. Natl. Acad. Sci. USA* **99,** 12287–12292.
Wu, Z., Bensinger, S. J., Zhang, J., Chen, C., Yuan, X., Huang, X., Markmann, J. F., Kassaee, A., Rosengard, B. R., Hancock, W. W., Sayegh, M. H., and Turka, L. A. (2004). Homeostatic proliferation is a barrier to transplantation tolerance. *Nat. Med.* **10,** 87–92.
Yamanouchi, J., Rainbow, D., Serra, P., Howlett, S., Hunter, K., Garner, V. E., Gonzalez-Munoz, A., Clark, J., Veijola, R., Cubbon, R., Chen, S. L., Rosa, R., *et al.* (2007). Interleukin-2 gene variation impairs regulatory T cell function and causes autoimmunity. *Nat. Genet.* **39,** 329–337.
You, S., Chen, C., Lee, W. H., Brusko, T., Atkinson, M., and Liu, C. P. (2004). Presence of diabetes-inhibiting, glutamic acid decarboxylase-specific, IL-10-dependent, regulatory T cells in naive nonobese diabetic mice. *J. Immunol.* **173,** 6777–6785.
You, S., Belghith, M., Cobbold, S., Alyanakian, M. A., Gouarin, C., Barriot, S., Garcia, C., Waldmann, H., Bach, J. F., and Chatenoud, L. (2005). Autoimmune diabetes onset results from qualitative rather than quantitative age-dependent changes in pathogenic T-cells. *Diabetes* **54,** 1415–1422.
Zacher, T., Knerr, I., Rascher, W., Kalden, J. R., and Wassmuth, R. (2002). Characterization of monocyte-derived dendritic cells in recent-onset diabetes mellitus type 1. *Clin. Immunol.* **105,** 17–24.
Zucchelli, S., Holler, P., Yamagata, T., Roy, M., Benoist, C., and Mathis, D. (2005). Defective central tolerance induction in NOD mice: Genomics and genetics. *Immunity* **22,** 385–396.

CHAPTER 6

Gene–Gene Interactions in the NOD Mouse Model of Type 1 Diabetes

**William M. Ridgway,* Laurence B. Peterson,[†]
John A. Todd,[‡] Dan B. Rainbow,[‡] Barry Healy,[‡]
Oliver S. Burren,[‡] and Linda S. Wicker[‡]**

Contents		
	1. Type 1 Diabetes is a Multigenic Disease	152
	2. Insights into a Multigenic Disease: The NOD Mouse	153
	3. Defining the Function of Alleles Altering the Frequency of T1D	161
	4. Functional Studies with NOD Congenic Mice: The Combination of Protective Alleles at *Idd3* and *Idd5* Prevent Diabetes and Insulitis	162
	5. Review of *Idd3*	163
	6. Review of *Idd5*	165
	7. Genetic Complexity of *Idd5*: Genetic Masking and the Discovery of *Idd5.3* and *Idd5.4*	167
	8. Combining Protective Alleles at *Il2/Idd3* with those from the *Idd5* Subregions	168
	9. Conclusion	169
	Acknowledgments	170
	References	170

Abstract   Human genome wide association studies (GWAS) have recently
identified at least four new, non-MHC-linked candidate genes or

* University of Pittsburgh School of Medicine, 725 SBST, Pittsburgh, Pennsylvania
[†] Merck Research Laboratories, Rahway, New Jersey
[‡] Juvenile Diabetes Research Foundation/Wellcome Trust Diabetes and Inflammation Laboratory, Cambridge Institute for Medical Research, Wellcome Trust/MRC Building, Addenbrooke's Hospital, Cambridge CB2 0XY, United Kingdom

Advances in Immunology, Volume 100   © 2008 Elsevier Inc.
ISSN 0065-2776, DOI: 10.1016/S0065-2776(08)00806-7   All rights reserved.

gene regions causing type one diabetes (T1D), highlighting the need for functional models to investigate how susceptibility alleles at multiple common genes interact to mediate disease. Progress in localizing genes in congenic strains of the nonobese diabetic (NOD) mouse has allowed the reproducible testing of gene functions and gene–gene interactions that can be reflected biologically as intra-pathway interactions, for example, IL-2 and its receptor CD25, pathway-pathway interactions such as two signaling pathways within a cell, or cell-cell interactions. Recent studies have identified likely causal genes in two congenic intervals associated with T1D, *Idd3*, and *Idd5*, and have documented the occurrence of gene–gene interactions, including "genetic masking", involving the genes encoding the critical immune molecules IL-2 and CTLA-4. The demonstration of gene–gene interactions in congenic mouse models of T1D has major implications for the understanding of human T1D since such biological interactions are highly likely to exist for human T1D genes. Although it is difficult to detect most gene–gene interactions in a population in which susceptibility and protective alleles at many loci are randomly segregating, their existence as revealed in congenic mice reinforces the hypothesis that T1D alleles can have strong biological effects and that such genes highlight pathways to consider as targets for immune intervention.

1. TYPE 1 DIABETES IS A MULTIGENIC DISEASE

The sequencing of the human and mouse genomes, followed by the successful utilization of single nucleotide polymorphisms (SNPs) for genome wide association studies (GWAS) in humans, has facilitated the identification of many new candidate genes in a variety of autoimmune diseases including type one diabetes (T1D) (Todd *et al.*, 2007; WTCCC, 2007). The candidate disease-causing alleles are functional variants commonly found in the population; they are not striking gain or loss of function mutations such as those characterized in the autoimmune regulator (AIRE) gene that cause autoimmune polyendocrine syndrome type I, a monogenic disease affecting endocrine glands and other organs (Anderson *et al.*, 2002; Manolio *et al.*, 2008). Instead, the apparent effect of any particular candidate gene on disease as measured in population-based studies is usually very low when compared to classic Mendelian loss of function mutations; most common variants discovered in GWAS have an odds ratio of less than 2.0 (Manolio *et al.*, 2008). (Odds ratio refers to the risk of disease in the presence of a disease susceptibility gene divided by the risk in the presence of a protective allele, e.g., an odds ratio of two means someone with the disease-associated SNP has a 2-fold increased risk of disease compared to someone with the alternative allele that is associated with protection from disease.) In the case of T1D, genetic studies have identified nine non-MHC genes or genetic regions associated with disease (Todd et al., 2007; WTCCC, 2007); most of these have odds ratios of less than 2. Such odds ratios contrast greatly with those of the peptide-binding class II molecules encoded within the MHC, with the most susceptible haplotypes having odds ratios greater than 10.0 (Cucca et al., 2001). Variation in MHC class II molecules is the major genetic component of T1D in both humans and the nonobese diabetic (NOD) mouse, a multigenic model of type 1 diabetes (Wicker et al., 1995). However, having one or more susceptibility alleles at the MHC is not sufficient to develop T1D, thereby emphasizing the importance of characterizing the non-MHC genes.

There has been a striking correspondence between the identified human T1D genes with insulin dependent diabetes (*Idd*) loci defined in NOD mice or with genetic manipulations carried out in the NOD model to mimic the human disease-causing genes. The peptide-binding class I and class II MHC molecules, CTLA-4, the IL-2/CD25 pathway, and insulin (both as an autoantigen and in regards to its thymic expression, which is important in deleting insulin-specific T cells) have all been implicated in both humans and mice (Rainbow et al., 2008; Wicker et al., 2005). This remarkable correspondence of the genetic basis of disease is a strong validation that immune pathways are involved in T1D pathogenesis and provides a strong rationale for continuing to study T1D in the NOD mouse model, especially as it applies to understanding the contribution of common gene variants. Moreover, some of the human genes identified as strong candidates do not have known functions, thereby providing the opportunity to discover how they contribute to T1D pathogenesis by the direct alteration of such genes in NOD mice. Thus, studies of human T1D genes in the NOD model should provide insights into autoimmunity via increased knowledge of both novel and established pathways.

2. INSIGHTS INTO A MULTIGENIC DISEASE: THE NOD MOUSE

Some have questioned the rationale for continuing efforts to discover and characterize common gene variants such as those discovered in human T1D since they have such small individual effects on disease, i.e., nearly all susceptibility alleles have an increased disease risk of less than 2-fold across the whole population. However, the small effect shown in human population studies does not mean these genes do not play major roles in the pathogenesis of T1D. The importance of discovering genes with apparently small effects in the outbred human population is shown by the large effects similar gene variants have on the frequency of T1D in congenic NOD mice. To illustrate how linkage studies and the subsequent development and analysis of congenic mice have facilitated the understanding of a multigenic disease, let us consider the insights gained from the linkage studies used to define B10-derived alleles providing protection from T1D (Ghosh *et al.*, 1993; Todd *et al.*, 1991; Wicker *et al.*, 1989).

The first linkage analysis showed that although homozygosity at the NOD MHC was required for a high penetrance of diabetes, it was not sufficient, non-MHC genes were also required to develop T1D (Wicker *et al.*, 1987). Another important finding was that T1D developed with a lower penetrance in MHC heterozygous mice demonstrating that the contribution of MHC-linked susceptibility alleles is dose-dependent (Wicker *et al.*, 1987, 1989). The MHC's critical effect on the genetic control of T1D pathogenesis has since been defined further in both humans and mice; gene variants encoding both class I and class II MHC molecules contribute to T1D susceptibility in both species, which emphasizes the importance of both CD4 and CD8 T cells in the destruction of beta cells within the islets (Hamilton-Williams *et al.*, 2001; Nejentsev *et al.*, 2007; Singer *et al.*, 1993; Slattery *et al.*, 1990).

The first large-scale linkage study used NOD mice and a B10 congenic strain having a NOD MHC region in order to perform the segregation analysis with full MHC-associated susceptibility in all progeny (Table 6.1) (Ghosh *et al.*, 1993; Todd *et al.*, 1991). Fourteen percent of the backcross 1 generation (to NOD) developed T1D and nine regions causing T1D or insulitis were identified, certainly more than had been expected assuming standard recessive susceptibility models (Ghosh *et al.*, 1993; Todd *et al.*, 1991). Although B10-derived alleles at many of the T1D regions provided potent protection from disease, another important lesson learnt was that some of the B10-derived regions conferred increased susceptibility. This finding, which was quite unexpected at the time, led to the hypothesis that each inbred strain of mouse, and by analogy each individual in the outbred human population, has a unique combination of alleles at genes that influence the immune system; some alleles will provide susceptibility to T1D, others resistance, and others will be neutral.

The other major lesson from the backcross 1 generation was that susceptibility alleles at any particular T1D region were not required for disease to develop, nor did any protective allele always prevent T1D. Therefore, none of the non-MHC genes affected T1D in a dominant or recessive manner, and no gene interactions were observed. Mathematical modeling of type 1 diabetes in both humans and mice suggests that a multiplicative model in which there is statistical independence between genotypes at the contributing loci best describes the observed genetic contributions, however, as discussed below, specific gene interactions are difficult to discern in a segregating population (Cordell *et al.*, 2001; Risch *et al.*, 1993; Todd *et al.*, 2007). It is also instructive to compare the

TABLE 6.1 Summary of the genetic control of T1D in the NOD mouse model

Strain	Diabetes	Interpretation
NOD (H-2^{g7})	80% in females	
B10.H-2^{g7}	None	Homozygous expression of the NOD-derived MHC is not sufficient to cause diabetes. Non-MHC NOD-derived genes are also required.
NOD × B10.H-2^{g7} F1	None	Homozygous expression of the NOD-derived MHC and one dose of all NOD-derived (or B10-derived) non-MHC susceptibility alleles are not sufficient to cause diabetes. B10-derived, non-MHC protective alleles have an overall dominant protective effect on the T1D phenotype.
(NOD × B10.H-2^{g7} F1) × NOD backcross one generation	14% in females	Homozygous expression of the NOD-derived MHC and a sufficient number of NOD-derived (or B10-derived) non-MHC susceptibility alleles are present in 14% of backcross one females. At least nine non-MHC regions influence T1D susceptibility in this backcross one generation.
(NOD × B10.H-2^{g7} F1) × (NOD × B10.H-2^{g7} F1) F2 generation	0.4% in females	Despite the homozygous expression of the NOD-derived MHC and, on average, 25% of *Idd* loci homozygous for susceptibility alleles each F2 mouse, only 0.4% of mice develop T1D. These results also support the hypothesis that in combination B10-derived, non-MHC protective alleles have an overall dominant protective effect on the T1D phenotype.

frequency of human T1D (0.1–0.5% in European populations, substantially less in non-European populations) with that in the (NOD × B10. H-2^{g7}) F2 generation, in which only 0.4% of F2 females developed T1D (Table 6.1) (McAleer et al., 1995). We now understand from detailed gene mapping studies and the analysis of NOD congenic mouse strains that a minimum of 20 non-MHC T1D regions differ between NOD and B10/B6 mice (Fig. 6.1 and Table 6.2). Additional *Idd* regions have been defined using other non-NOD strains for linkage studies or to develop congenic strains. In the F2 generation these 20 regions segregate randomly and in order to develop diabetes, a minimum number of susceptibility alleles must be inherited, an event occurring in approximately 1/200 mice despite the fact that each inherited two doses of the NOD MHC region (Table 6.1). The B10 strain differs extensively throughout its genome from the NOD strain, and using it maximized the representation of functional alleles in the F2 generation, thereby modeling the randomly segregating human population. The simultaneous segregation of a large number of human alleles also explains the need for large sample sizes, i.e., thousands of matched cases and controls, to detect the effects of individual alleles.

The low incidence of T1D in the mouse F2 generation and in the outbred human population indicates that on balance there are more protective than susceptibility alleles at most genes in the population and/or that protective alleles (or the resulting protective cellular mechanisms modulated by the genes) are dominant. Although it is possible that particular protective alleles "mask" specific susceptibility alleles, or vice versa, it is difficult in an outbred population to have sufficient sample sizes to test such hypotheses adequately. Such specific masking effects, also termed epistasis, are readily observed in congenic mouse studies as detailed below. Until we understand the cellular mechanisms that mediate susceptibility and resistance by individual T1D alleles, we can only speculate on the question of whether masking is caused by the interaction of multiple, independent pathways, each having variants at multiple genes, or that gene variants interact within a limited number of pathways, all of which contribute to T1D pathogenesis in most individuals.

The remainder of this review focuses on how, within the context of NOD congenic strains, the effects of individual protective and susceptible alleles can be studied. Not only can T1D frequency and the development of autoantibodies and insulitis be monitored, cell transfer experiments can be performed to determine the cellular basis for an altered immune response. Ultimately, the analysis of variants, or surrogate variants, of common human T1D genes will be testable in the NOD mouse model or in related congenic strains, similar to previous studies testing biological hypotheses concerning T1D-modulating variants of human class I and class II MHC molecules and human insulin.

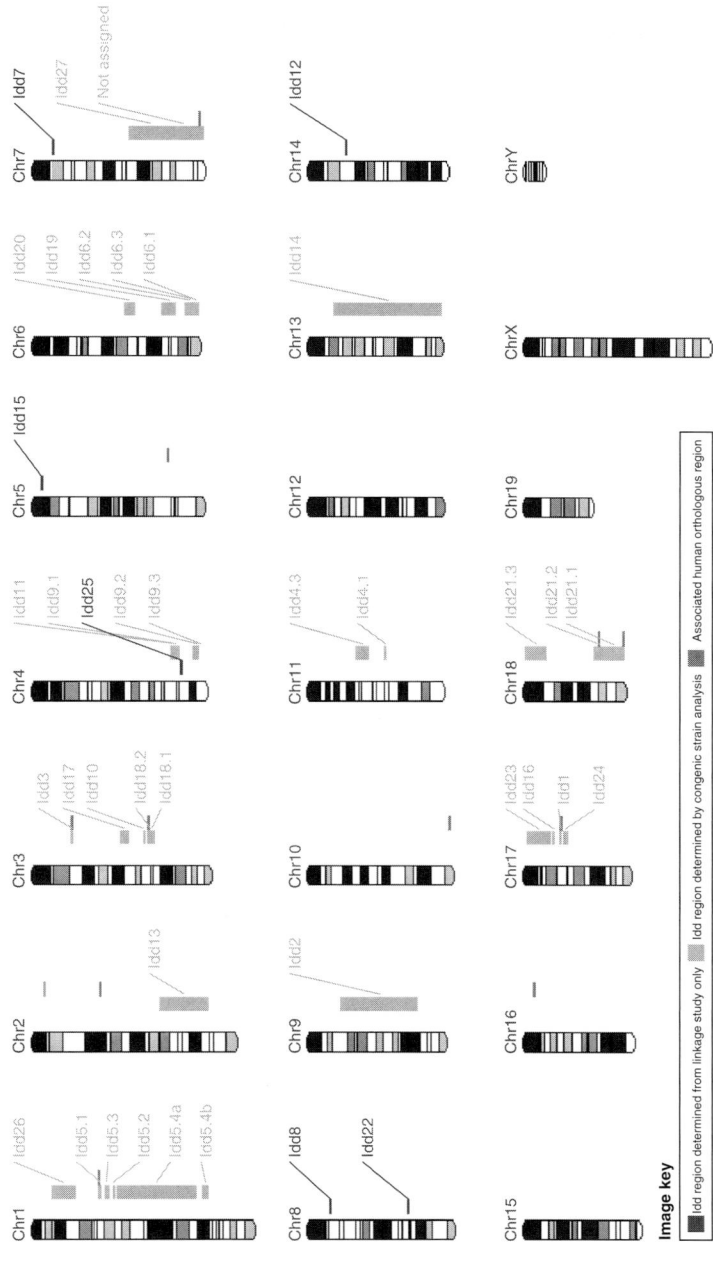

FIGURE 6.1 The *Idd* regions. Table 6.2 details the location and size (in the case of congenic intervals) of the gene regions depicted in Fig. 6.1 affecting the frequency of diabetes in NOD mice. Resistant (R) and Susceptible (S) refers to the diabetes phenotype associated with the *Idd* allele present in the non-NOD strain. The *Idd1* interval is defined by *H2-K1* at the proximal end and *H2-D1* at the distal end and includes the class II region. Mouse Ensembl release 50 was used to determine the size and position of the *Idd* regions. The figure was taken from http://www.t1dbase.org/page/PosterView/MouseIddRegions. Candidate genes for each *Idd* region are also listed at this web site.

TABLE 6.2 The *Idd* regions

Idd Region	Non-NOD strain	Non-NOD strain T1D phenotype	Chr	Limits of congenic interval or peak of linkage (Mb)	Interval size Mb	Reference
Idd1	B10	R	17	34.133–35.405	1.271*	Reviewed in Wicker et al. (1995)
Idd2	B10	R	9	32.308–98.698	66.39	Pearce (1998)
Idd3	B6	R	3	36.627–37.277	0.651	Yamanouchi et al. (2007)
Idd4.1	NOR	R	11	69.76–71.152	1.392	Ivakine et al. (2006)
Idd4.3	C57L	R	11	44.553–55.855	11.303	Litherland et al. (2005)
Idd5.1	B10	R	1	60.833–62.840	2.007	Wicker et al. (2004)
Idd5.2	B10	R	1	73.984–75.465	1.481	Wicker et al. (2004)
Idd5.3	B10	R	1	66.530–70.084	3.554	Hunter et al. (2007)
Idd5.4a	B10	S	1	77.143–147.307	70.164	Hunter et al. (2007)
Idd5.4b	B10	S	1	152.632–157938	5.306	Hunter et al. (2007)
Idd6.1	C3H	R	6	146.378–149.517	3.140	Hung et al. (2006)
Idd6.2	C3H	R	6	143.560–146.378	2.818	Hung et al. (2006)
Idd6.2	B6	R	6	137.404–146.386	8.982	Bergman et al. (2003)
Idd6.3	C3H	R	6	146.262–147.388	1.126	Hung et al. (2006)
Idd7	B6	S	7	Peak between 21.0–43.0		Serreze et al. (2008)
Idd7	B10	S	7	Peak at 19.997		Ghosh et al. (1993)
Idd7	NON	S	7	Peak at 19.997		McAleer et al. (1995)
Idd8	B10	S	14	Peak at 21.66		Ghosh et al. (1993); Liston et al. (2004)

(*continued*)

TABLE 6.2 (continued)

Idd Region	Non-NOD strain	Non-NOD strain T1D phenotype	Chr	Limits of congenic interval or peak of linkage (Mb)	Interval size Mb	Reference
Idd9.1	B10	R	4	128.365–131.179	2.813	Lyons et al. (2000b), Wicker et al., unpublished
Idd9.2	B10	R	4	144.968–149.098	4.13	Siegmund et al. (2000), Wicker et al., unpublished
Idd9.3	B10	R	4	149.300–150.522	1.222	Cannons et al. (2005)
Idd10	B6	R	3	99.699–100.577	0.877	Penha-Goncalves et al. (2003), Wicker et al., unpublished
Idd11	B6	R	4	125.017–132.983	7.966	Brodnicki et al. (2005)
Idd12	B6	R	14	Peak at 35.170		Liston et al. (2004), Morahan et al. (1994)
Idd13	NOR	R	2	114.118–158.330	44.212	Chen et al. (2007), Serreze et al. (1998)
Idd13—B2m	NOR	R	2	114.118–130.275	16.157	Chen et al. (2007)
Idd13—not B2m	NOR	R	2	121.973–134.812	12.839	Chen et al. (2007)
Idd14	B6	S	13	25.424—120.284	94.860	Brodnicki et al. (2003)
Idd15	NON	R	5	Peak at 8.798		McAleer et al. (1995)
Idd16	B6	R	17	26.318–29.405	3.086	Deruytter et al. (2004)
Idd17	B6	R	3	79.484–87.106	7.621	Podolin et al. (1997)

(continued)

TABLE 6.2 (continued)

Idd Region	Non-NOD strain	Non-NOD strain T1D phenotype	Chr	Limits of congenic interval or peak of linkage (Mb)	Interval size Mb	Reference
Idd18.1	B6	R	3	108.993–109.597	0.604	Lyons et al. (2001), Wicker et al., unpublished
Idd18.2	B6	S	3	100.95–108.222	7.272	Lyons et al. (2001), Wicker et al., unpublished
Idd19	C3H	S	6	117.439–128.468	11.029	Morin et al. (2006), Rogner et al. (2001)
Idd20	C3H	R	6	83.595–91.990	8.394	Morin et al. (2006)
Idd21.1	ABH	R	18	69.192–90.722	21.530	Hollis-Moffatt et al. (2005)
Idd21.2	ABH	S	18	64.618–74.588	9.970	Hollis-Moffatt et al. (2005)
Idd21.3	ABH	R	18	0–21.671	21.672	Hollis-Moffatt et al. (2005)
Idd22	ALR	R	8	Peak at 90.626		Mathews et al. (2003)
Idd23	B6	R	17	3.924–26.318	22.394	Deruytter et al. (2004)
Idd24	B6	R	17	35.340–44.938	9.598	Deruytter et al. (2004)
Idd25	NOR	R	4	Peak at 133.341		Reifsnyder et al. (2005)
Idd26	NOR	R	1	19.802–40.319	20.517	Reifsnyder et al. (2005)
Idd27	CBA	R	7	86.521–127.029	40.508	Chen et al. (2005)
Not assigned	C57L	R	7	117.936–152.524	34.588	Chen et al. (2005)

This table details the location and size (in the case of congenic intervals) of the gene regions depicted in Fig. 6.1 affecting the frequency of diabetes in NOD mice. Resistant (R) and Susceptible (S) refers to the diabetes phenotype associated with the *Idd* allele present in the non-NOD strain

* The *Idd1* interval is defined by *H2-K1* at the proximal end and *H2-D1* at the distal end and includes the class II region. Mouse Ensembl release 50 was used to determine the size and position of the *Idd* regions. The figure was taken from http://www.t1dbase.org/page/PosterView/MouseIddRegions. Candidate genes for each *Idd* region are also listed at this web site.

3. DEFINING THE FUNCTION OF ALLELES ALTERING THE FREQUENCY OF T1D

The overarching goal is to define how, at both the molecular and cellular levels, each human T1D gene variant alters disease pathogenesis. Each disease-associated gene presents a unique challenge and many details can only be approached via animal modeling. In regard to the molecular basis of disease alleles, some variants are relatively straightforward, i.e., those causing an amino acid change (or changes) that measurably alters a known function of the protein, such as variants in the peptide binding pockets of class I and class II molecules or the amino acid-changing SNP in *PTPN22* that alters its enzyme function. However, some amino acid changes occur in proteins without known functions or cause subtle functional alterations that are difficult to detect or are only detectable under certain stimulation conditions. Other amino acid substitutions can alter post-translational modifications by modifying glycosylation or cellular localization motifs. A SNP causing an amino acid change may not alter protein function but instead can alter splicing preferences or efficiencies since exons contain exonic splicing enhancer (ESE) and suppressor or silencer (ESS) motifs (Zheng, 2004). The molecular basis of noncoding SNPs associated with T1D is often even more difficult to define since small differences in mRNA or protein production are challenging to detect reproducibly. One example is the ~2-fold difference of IL-2 produced by polyclonal CD4 and CD8 T cells isolated from mice having susceptibility and resistance alleles (Yamanouchi *et al.*, 2007). In other examples, nucleotide variation in the CTLA-4 gene can produce differing ratios of mRNAs arising from alternative splicing in both humans and NOD mice (Ueda *et al.*, 2003) and a SNP in the human IL-7R gene is associated with an altered ratio of the alternatively spliced transmembrane and soluble isoforms (Gregory *et al.*, 2007).

The cellular basis of T1D gene action can be approached for both human and mouse T1D genes by determining which cell types express the gene, with and without appropriate stimulation protocols. Unfortunately many variant genes are expressed in most immune cells and some T1D genes are expressed in all cells, thereby limiting the usefulness of the information in regard to defining the cellular basis of disease causation. Transfer of bone marrow and defined cell populations derived from spleen and lymph nodes cells have allowed delineation of the cell types in which some *Idd* loci function, but of course such studies are limited to mouse models. For example, NOD.$Ig\mu^{null}$ mice that are irradiated and reconstituted with syngeneic bone marrow plus NOD B cells develop T1D. However, if the NOD B cells are replaced with B cells purified from *Idd5* or *Idd9/11* mice, the mice are protected from T1D (Silveira et al., 2006). These results do not imply that B cells are the only cell type in which T1D-resistance alleles at *Idd9/11* or *Idd5* function to prevent diabetes at a cellular level. However, the observations do imply that the gene products encoded in these *Idd* regions alter B cells. It is also important to remember in such studies that the transferred B cells are developed in an environment in which all cell types had the protective alleles at *Idd9/11* or *Idd5*. Although potential caveats abound in transfer studies, they remain important tools for defining in which cell types T1D genes mediate their disease-causing functions.

4. FUNCTIONAL STUDIES WITH NOD CONGENIC MICE: THE COMBINATION OF PROTECTIVE ALLELES AT *IDD3* AND *IDD5* PREVENT DIABETES AND INSULITIS

Whereas the effects of T1D genes are relatively small in a segregating population, in congenic strains of mice where the NOD susceptibility genes are held constant except for the region under evaluation, easily measurable effects on the frequency of T1D have been observed with single protective genes. Mapping of a T1D gene by reiteratively reducing the size of the congenic interval has in many cases, revealed that regions once thought to have a single *Idd* gene harbor multiple, closely linked T1D genes (Hunter *et al.*, 2007; Podolin *et al.*, 1998; Wicker *et al.*, 2004). The earliest efforts to define *Idd* genes by congenic mapping included the *Idd3* and *Idd5* regions, which were known to include the IL-2 and CTLA-4 genes respectively. Protective alleles at *Idd3* alone on the NOD background result in 20% diabetes as compared to 80% in the parental NOD strain; *Idd5* alone results in 40% disease. Because of the candidacies of IL-2 and CTLA-4 genes, both critical for the maintenance of immune homeostasis, we hypothesized that the combination of protective alleles in the *Idd3* and *Idd5* region in an *Idd3/Idd5* double congenic strain could potentially provide a high degree of protection. This prediction proved to be correct since nearly complete protection from diabetes (only ~1% of double congenic female mice develop T1D), insulitis, and insulin autoantibodies was observed (Hill *et al.*, 2000; Robles *et al.*, 2003). Cordell *et al.* (2001) modeled the genetic interactions of protective alleles at *Idd3* and *Idd5* as well as those of *Idd3* and *Idd10* and concluded that synergistic gene–gene interactions in these two combinations of congenic intervals were evident. Since genetic interactions were not observed in the (NOD × B10.H-2^{g7}) × NOD backcross segregation analysis detailed earlier in this review, the conclusion that genetic interactions could be observed in congenic mice led us to hypothesize that the simultaneous segregation of alleles at multiple *Idd* loci in backcross mice prevented the detection of individual gene–gene interactions. The status of the T1D-causing biological pathways in each mouse in the backcross population would be influenced by the combination of all protective and susceptible *Idd* alleles inherited; therefore, the effect of alleles at a particular *Idd* locus would be more variable than in a congenic strain comparison where all other loci have the same alleles.

This *Idd3* and *Idd5* combination of protective alleles has also been shown to resist autoimmunity robustly in other experimental situations. For example, the Mathis group demonstrated that breeding an AIRE knockout onto the B6 background resulted in minimal autoimmunity, but breeding it onto the NOD background resulted in severe autoimmunity in multiple organs (Jiang *et al.*, 2005). Nonetheless, the combination of *Idd3* and *Idd5* abrogated the combined effect of the AIRE knockout and the non-*Idd3/Idd5* NOD background genes (Jiang *et al.*, 2005). The *Idd3/Idd5* protective alleles together were shown to abrogate salivary gland dysfunction in NOD.$H2^b$ mice and B6 mice congenic for the NOD alleles at *Idd3* and *Idd5* demonstrated autoimmunity in the salivary and lacrimal glands including decreased secretory flow rates in both tissues equivalent to those seen in the NOD strain (Cha *et al.*, 2002). Mechanistic studies by Martinez *et al.* demonstrated that *Idd3/Idd5* protective alleles enabled elimination of pathogenic CD8 T cells in the pancreatic lymph nodes following activation by autoantigen (Martinez *et al.*, 2005). Liston *et al.* linked defective NOD negative selection to the *Idd5* region (Liston *et al.*, 2004), although the combined effect of *Idd3* and *Idd5* alleles on this phenotype was not tested. The linkage of NOD-derived *Idd5* alleles to defective thymic apoptosis was confirmed by Zuchelli *et al.* in fetal thymic organ culture (Zucchelli *et al.*, 2005).

5. REVIEW OF *IDD3*

The *Idd3* region was first implicated in a T1D linkage analysis of (NOD × B10.$H2^{g7}$)F1 mice backcrossed to the NOD parental strain (Ghosh *et al.*, 1993; Todd *et al.*, 1991). The linkage was confirmed by the construction of an *Idd3* congenic strain which demonstrated protection from diabetes; *Il2* was within the congenic interval (Wicker *et al.*, 1994). Whilst *Il2*, which is located on chromosome 3 at 37.3 Mb, was an attractive candidate from the first linkage analysis, subsequent work involved continued narrowing of the interval containing *Idd3* via the selection of recombination events and selective breeding and the development of new congenic strains (Lyons and Wicker 1999). The congenic strain positional cloning approach reduced the *Idd3* interval to 650 kb, a region which includes *Il2*, *Il21*, and several other genes (Denny *et al.*, 1997; Lyons *et al.*, 2000a; Yamanouchi *et al.*, 2007). Given IL-2's critical role in the immune system, this lymphokine was an attractive candidate gene in a relatively small interval. Proving that IL-2 was in fact *Idd3*, however, was a prolonged process that illustrates important principles in defining functional alleles of the immune system (Denny *et al.*, 1997; Lyons *et al.*, 2000a; Podolin *et al.*, 2000). An initial haplotype analysis, which was conducted by determining if *Idd3* regions of additional inbred strains have protective or susceptibility alleles and performing a sequence comparison of disease-resistant versus disease-susceptible alleles, pointed to the possibility that amino acid changes not affecting the binding of IL-2 to its high affinity receptor but altering its glycosylation, were responsible for the functional allelic difference. Although the identity of the *Idd3* gene remained unresolved, evidence of its functional role in multiple immune phenotypes accumulated. The NOD *Idd3* allele was shown to play a role in transplantation tolerance; islet allograft survival was prolonged in diabetes-resistant NOD.B6 *Idd3* mice compared to NOD mice and shortened in B6 mice congenic for the NOD *Idd3* allele as compared to B6 mice. Thus, the *Idd3* locus affects both self and allo-tolerance and the allelic difference is apparent on a diabetes-free background (B6) (Pearson *et al.*, 2004). The B6 *Idd3* locus also partially restored CD8 T cell tolerance assessed in the pancreatic lymph nodes (Martinez *et al.*, 2005). Finally, *Idd3* was shown to play a broader role in autoimmune disease, since mice with a B6 *Idd3* locus in conjunction with other *Idd* resistance loci had no diabetes, but developed autoimmune biliary disease (Irie *et al.*, 2006; Koarada *et al.*, 2004). This role in a broad spectrum of autoimmune diseases was supported by evidence implicating the *Idd3* region in experimental autoimmune encephalomyelitis (EAE) and in neonatal thymectomy-induced autoimmune ovarian dysgenesis (Encinas *et al.*, 1999; Teuscher *et al.*, 1996).

Yamanouchi *et al.* provided evidence that the *Idd3* gene is *Il2* by showing that the B6 allele produces approximately 2-fold more IL-2 mRNA than the NOD allele (Yamanouchi *et al.*, 2007). No allelic-specific production of IL-21 was seen in this study. An extensive haplotype analysis and sequencing effort revealed a large number of candidate SNPs throughout the IL-2 gene including the 8 kb upstream region that is known to be essential for the appropriate *in vivo* regulation of the IL-2 gene (Yui *et al.*, 2004). Results from the haplotype analysis disproved the hypothesis that IL-2 glycosylation differences contribute to the molecular basis of *Idd3* and the authors proposed long-range chromatin remodeling of the IL-2 gene as the molecular basis of the observed allelic-specific transcription differences. Perhaps the most convincing evidence that *Idd3* is explained by a relatively small change in IL-2 production is that the frequency of T1D increased when IL-2 production was decreased by 50% in hemizygous IL-2 KO NOD mice that have only one functional allele of *Il2*, not two (Yamanouchi *et al.*, 2007).

Establishing that *Il2* is the causative gene in *Idd3* raises the next question, which is essential for realizing eventual therapeutic results:

what are the cellular mechanisms mediating *Idd3*'s ability to alter T1D pathogenesis? A major issue in identifying disease-causing pathways or molecules acting in the immune system is that many of them function in various cell types, some of which have opposing actions in the development of tolerance or immunity. In the Yamanouchi study, enhanced IL2 production correlated with protection from T1D despite the fact that increased production of IL-2 by islet-specific CD8 T cells caused them to differentiate more rapidly into cytotoxic effector cells (Yamanouchi *et al.*, 2007). The hypothesis put forward to explain this observation was that increased IL-2 production by pathogenic CD8 T cells increased the effectiveness of the CD4+ CD25+ Treg population, a cell subset that requires IL-2 for its maintenance (Gavin *et al.*, 2007). In the context of a balanced immune response wherein IL-2 enhances both effector T cells and regulatory cells, the results from Yamanouchi *et al.* support the hypothesis that regulation is dominant.

6. REVIEW OF *IDD5*

A linkage study identifying *Idd5* on proximal chromosome 1 (Cornall *et al.*, 1991) was confirmed by the development of a NOD.B10 *Idd5* congenic strain (Hill *et al.*, 2000). Hill *et al.* demonstrated that *Idd5* was actually at least two genes: *Idd5.1*, which included the candidate genes *Ctla4*, *Cd28* and *Icos*, and *Idd5.2* having the candidate gene *Nramp1*, subsequently renamed to *Slc11a1* (Hill *et al.*, 2000). Subsequent novel congenic strains reduced the *Idd5.1* region to a 2.1 Mb region that contained four genes including *Ctla4* and *Icos*; *Cd28* was excluded (Wicker *et al.*, 2004). *Idd5.2* was narrowed to a 1.5 Mb region which contains at least 45 genes and still includes *Slc11a1* (Wicker *et al.*, 2004) (also see Table 6.3).

The subsequent characterization of *Idd5* illustrates some of the challenges and complexities of proving the identity of a causal gene in an *Idd* interval once that region has been reduced in size as much as is practically possible (to less than approximately 2 Mb). The difficulties were different for each *Idd5* subregion. For example, *Idd5.1* had only a few genes, but three of them were known to have significant effects on the immune system (*Ctla4*, *Cd28*, and *Icos*). Moreover, *Ctla4* and *Icos* are very closely linked on chromosome 1, making exclusion of one or the other by random recombination unlikely. Finally, both *Ctla4* and *Icos* expression varied between T cells obtained from NOD and NOD.B10 *Idd5.1* congenic mice (Greve *et al.*, 2004). A recently completed haplotype analysis similar to that performed for the *Idd3* region detailed above supports the candidacy of *Ctla4* as *Idd5.1* (Wicker *et al.*, submitted for publication).

The proposed molecular basis of *Ctla4* variation reflected in the *Idd5.1* alleles is a SNP in exon 2 of *Ctla4* affecting the levels of a newly discovered

TABLE 6.3 Subregions of *Idd5* on chromosome 1

Region	Start of congenic region on chromosome 1, size of interval	
Idd5.1	60.9 Mb, 2.1 Mb	*Ctla4* is the primary candidate gene. Protective alleles at *Idd5.1/Ctla4* are dominant over susceptibility alleles at *Idd5.4*.
Idd5.3	66.3 Mb, 3.5 Mb	*Acadl* is the primary candidate gene. Susceptibility alleles at *Idd5.4* mask protective alleles at *Idd5.3* in some genetic contexts.
Idd5.2	74.1 Mb, 1.5 Mb	*Slc11a1* (formerly called *Nramp1*) is the primary candidate gene. Susceptibility alleles at *Idd5.4* mask protective alleles at *Idd5.2* in some genetic contexts.
Idd5.4a *Idd5.4b*	77.1 Mb, 70.1 Mb 152.6 Mb, 5.3 Mb	Candidate gene definition requires that this interval be reduced in size by further congenic mapping analysis. *Idd5.4* is composed of a B10-derived segment containing a double recombination event that was fixed while the region was backcrossed to the NOD strain. The two B10-derived *Idd5.4* intervals are termed *Idd5.4a* and *Idd5.4b* (see Fig. 6.1). *Idd5.4*-mediated susceptibility to T1D is masked by protective alleles at *Idd5.1/Ctla4*.

(at the time) isoform of CTLA-4 (Ueda *et al.*, 2003; Wicker *et al.*, 2004). The novel, alternatively spliced mRNA skips exon 2, which encodes the B7-1 binding domain and therefore its protein product was termed ligand-independent (li) CTLA-4 (Ueda *et al.*, 2003). The exon 2 SNP has been proposed to affect an exonic splicing silencer motif, with the stronger silencing motif present in the protective allele. Therefore, NOD T cells expressed significantly lower levels of liCTLA-4 than do T cells having the B10 allele at *Ctla4* (Ueda *et al.*, 2003). The liCTLA-4 isoform mediates negative signaling in T cells via binding and dephosphorylating the TcR zeta chain (Vijayakrishnan *et al.*, 2004).

As compared to *Idd5.1*, the *Idd5.2* region presented a different set of difficulties. Despite the B10-derived congenic interval having been reduced to 1.52 Mb, less than that of *Idd5.1*, *Idd5.2* contains 10-fold more genes than the *Idd5.1* interval. The clear functional variation in *Slc11a1* between the NOD and B6 alleles made it a tantalizing candidate gene from the earliest linkage studies (Cornall et al., 1991) and human *SLC11A1* has been associated with T1D and rheumatoid arthritis (Runstadler et al., 2005; Shaw et al., 1996). In macrophages, the Nramp1 protein facilitates the killing of certain intracellular pathogens; the NOD *Slc11a1* allele encodes a functional protein whereas the B6 allelic product has a loss of function mutation (Wicker et al., 2004). Kissler et al. directly tested the role of Nramp1 in diabetes pathogenesis using RNA interference hypothesizing that the natural mutation that prevents Nramp1 function could be mimicked by decreasing its expression using lentiviral transgenesis (Kissler et al., 2006). A decrease in T1D was observed and susceptibility to *Salmonella* infection was increased in mice having the knockdown of Nramp1 mRNA thereby strongly supporting the candidacy of *Slc11a1* as *Idd5.2*. It has been demonstrated recently that Nramp1 is expressed in dendritic cells and affects antigen presentation (Stober et al., 2007) providing additional possibilities of how *Slc11a1* might modulate the autoimmune response.

7. GENETIC COMPLEXITY OF *IDD5*: GENETIC MASKING AND THE DISCOVERY OF *IDD5.3* AND *IDD5.4*

The first set of congenic mapping results indicated that the *Idd5* region was composed of two closely linked genes, *Idd5.1* and *Idd5.2*, with *Idd5.1* being partially protective when isolated as a single *Idd* region. In contrast to *Idd5.1* congenic strains having a reduced frequency of T1D as compared to the parental NOD strain, an *Idd5.2* congenic strain did not have resistance to T1D in the absence of protective alleles at the linked *Idd5.1* gene (Hill et al., 2000).

The inability to reconstitute *Idd5*-mediated protection by reuniting isolated *Idd5.1* and *Idd5.2* protective alleles in a single strain (Hunter et al., 2007) resulted in the discovery of two additional gene regions on chromosome 1 affecting the frequency of diabetes in the study by Hill et al. (2000): a B10-derived protective allele at *Idd5.3*, which is located between *Idd5.1* and *Idd5.2*, and a B10-derived T1D *susceptibility* allele at *Idd5.4*, which is distal of *Idd5.2* (Table 6.3). The protective, B10-derived allele at *Idd5.3* is needed in addition to protective alleles at *Idd5.1* and *Idd5.2* to reconstitute the T1D frequency originally observed for mice congenic for the entire *Idd5* region (Hunter et al., 2007). *Idd5.3* is defined as a 3.5 Mb interval encompassing 11 genes. Using genome wide microarray analysis to assess mRNA expression in activated CD4+ T cells from NOD and NOD.B10 *Idd5.3* congenic mice, Irie *et al.* (2008) strongly implicated *Acadl* as a candidate gene in the *Idd5.3* interval since *Acadl* was the only highly differentially expressed gene in the relatively small *Idd5.3* interval. Interestingly, ACADL has no known immune functions but rather functions to control the first step in fatty acid β-oxidation (Matsubara *et al.*, 1989). Understanding the mechanism by which variation of a gene involved in energy production could potentially have such profound effects on T1D is an area of active research (Fox *et al.*, 2005).

The discovery of the B10-derived susceptibility allele at *Idd5.4* was quite serendipitous since protective versus susceptibility alleles at *Idd5.4* have no effect on the protection from T1D provided by the combination of protective alleles at all of the *Idd5* subregions, *Idd5.1, Idd5.2*, and *Idd5.3*. However, if the *Idd5.1/Ctla4* protective allele is replaced with a susceptibility allele (but continuing to have protective alleles at *Idd5.3* and *Idd5.2*) the status of the *Idd5.4* allele now has a dramatic influence on the frequency of T1D (Hunter *et al.*, 2007). Thus the effect of the *Idd5.4* susceptibility allele is masked by a protective allele at *Idd5.1/Ctla4*. In other words, when the B10-derived *Idd5.4* susceptibility allele is on the NOD background, it can mask the protective effects of at least certain genes (protective alleles at *Idd5.2* and *Idd5.3*) but the *Idd5.4* susceptibility allele in turn is masked by a protective allele at *Ctla4/Idd5.1*. These results imply that the *Idd5.4* pathway can in some manner be influenced by the status of negative signaling mediated by CTLA-4. Unfortunately, the *Idd5.4* region is ~75 Mb and additional congenic strains will be required to define candidate genes for this masking effect. It is quite sobering to realize how susceptible and protective *Idd* alleles linked on the same chromosome of inbred strains of mice can interact during the development of congenic strains and thus confound the interpretation of disease frequencies.

8. COMBINING PROTECTIVE ALLELES AT *IL2/IDD3* WITH THOSE FROM THE *IDD5* SUBREGIONS

As detailed earlier, Hill *et al.* had originally demonstrated that the combination of protective alleles at *Il2/Idd3* and *Idd5* led to virtually a complete elimination of both insulitis and diabetes (Hill *et al.*, 2000). Delineation of the direct effects of *Idd5.1/Ctla4* with *Il2/Idd3* however, produced a surprise: Hunter *et al.* found that *Idd3/Idd5.1* mice were not more protected from diabetes than *Idd3* mice (Hunter *et al.*, 2007). Another double congenic strain that was tested had protective alleles at *Idd5.1* and *Idd5.3* in addition to *Idd3*; this strain was equivalent to the *Idd3/Idd5* strain in regard to resistance to diabetes but insulitis was more prevalent (Hunter *et al.*, 2007).

Despite limited knowledge of the "causative" gene product in the *Idd5.3* and *5.4* intervals, the genetic results allow us to draw some conclusions about the immune pathways involved in T1D. As Hunter *et al.* point out, since liCTLA4 acts as a negative regulator of T cell function and its expression is increased by the protective allele, the B10 *Idd5.4* susceptibility allele could mediate an immune event that is counter-regulated by CTLA-4. The disease regulatory effect of CTLA-4 is dose dependent (Hunter *et al.*, 2007), suggesting a quantitative interaction between CTLA-4 and *Idd5.4* gene product at the cellular level. The pathway involving "masking" of the CTLA-4 effect on disease has direct relevance to human genetic studies of T1D, since it could explain why CTLA-4 is more strongly associated with a distinct subset of T1D patients who develop anti-thyroid autoantibodies early in life (Howson *et al.*, 2007).

This result demonstrates that the notion that any given genetic background can be described as "protective" against autoimmunity should be abandoned in favor of consideration of the effect of particular loci in the context of other loci. This point was also well demonstrated by the striking finding that the diabetogenic T cell clone BDC2.5, when placed as a TCR on the B6 background, had a much higher penetrance of diabetes than when placed on the NOD background (Gonzalez *et al.*, 1997). Finally, there is now at least one example of gene interaction in human T1D, one that involves *PTPN22* and HLA class II genotypes (Hermann *et al.*, 2006; Smyth *et al.*, 2008; Steck *et al.*, 2006). In this situation, the effect of a susceptibility allele at *PTPN22* is greater in lower-risk HLA class II genotypes than in the highest risk class II genotypes.

9. CONCLUSION

The identification of common genetic variants implicated in the autoimmune pathogenesis of T1D in NOD mice has progressed from linkage studies to initial insights into the molecular and cellular consequences of alleles mediating protection or susceptibility to disease. The identification of novel candidate genes from human genome-wide association studies in addition to those T1D genes already discovered underscores a common pathogenesis for human and mouse T1D and emphasizes the relevance of the NOD strain for modeling human variants. Using congenic mice, gene–gene interactions and gene masking effects have been observed that make large impacts on the T1D frequency whereas these effects are mostly hidden in a genetically segregating population such as a backcross one or an F2 generation, or in conventional genetic association studies in humans. We propose that the obfuscation of the genetic control of disease occurs because the T1D alleles that are randomly segregating each alter one or more T1D-causing molecular or cellular pathways, often resulting in a series of opposing effects on the expression of the T1D phenotype. Thus, although most human non-MHC susceptibility alleles confer an increased disease risk of less than 2-fold, and many less than 1.3-fold, the biological pathways affected by the human T1D genes are functional in the segregating population and should be evaluated for targeting by immune-based therapeutic strategies.

ACKNOWLEDGMENTS

WMR is funded by the NIH Autoimmunity Centers of Excellence. LSW and JAT are supported by grants from the Juvenile Diabetes Research Foundation (JDRF) and the Wellcome Trust and LSW is a JDRF/Wellcome Trust Principal Research Fellow. The Cambridge Institute for Medical Research is the recipient of a Wellcome Trust Strategic Award (079895).

REFERENCES

Anderson, M. S., Venanzi, E. S., Klein, L., Chen, Z., Berzins, S. P., Turley, S. J., von Boehmer, H., Bronson, R., Dierich, A., Benoist, C., and Mathis, D. (2002). Projection of an immunological self shadow within the thymus by the aire protein. *Science* **298,** 1395.

Bergman, M. L., Duarte, N., Campino, S., Lundholm, M., Motta, V., Lejon, K., Penha-Goncalves, C., and Holmberg, D. (2003). Diabetes protection and restoration of thymocyte apoptosis in NOD *Idd6* congenic strains. *Diabetes* **52,** 1677.

Brodnicki, T. C., Quirk, F., and Morahan, G. (2003). A susceptibility allele from a non-diabetes-prone mouse strain accelerates diabetes in NOD congenic mice. *Diabetes* **52,** 218.

Brodnicki, T. C., Fletcher, A. L., Pellicci, D. G., Berzins, S. P., McClive, P., Quirk, F., Webster, K. E., Scott, H. S., Boyd, R. L., Godfrey, D. I., and Morahan, G. (2005). Localization of *Idd11* is not associated with thymus and NKT cell abnormalities in NOD mice. *Diabetes* **54,** 3453.

Cannons, J. L., Chamberlain, G., Howson, J., Smink, L. J., Todd, J. A., Peterson, L. B., Wicker, L. S., and Watts, T. H. (2005). Genetic and functional association of the immune signaling molecule 4-1BB (CD137/TNFRSF9) with type 1 diabetes. *J. Autoimmun.* **25,** 13.

Cha, S., Nagashima, H., Brown, V. B., Peck, A. B., and Humphreys-Beher, M. G. (2002). Two NOD *Idd*-associated intervals contribute synergistically to the development of autoimmune exocrinopathy (Sjogren's syndrome) on a healthy murine background. *Arthritis Rheum.* **46,** 1390.

Chen, J., Reifsnyder, P. C., Scheuplein, F., Schott, W. H., Mileikovsky, M., Soodeen-Karamath, S., Nagy, A., Dosch, M. H., Ellis, J., Koch-Nolte, F., and Leiter, E. H. (2005). "Agouti NOD": Identification of a CBA-derived *Idd* locus on Chromosome 7 and its use for chimera production with NOD embryonic stem cells. *Mamm. Genome* **16,** 775.

Chen, Y. G., Driver, J. P., Silveira, P. A., and Serreze, D. V. (2007). Subcongenic analysis of genetic basis for impaired development of invariant NKT cells in NOD mice. *Immunogenetics* **59,** 705.

Cordell, H. J., Todd, J. A., Hill, N. J., Lord, C. J., Lyons, P. A., Peterson, L. B., Wicker, L. S., and Clayton, D. G. (2001). Statistical modeling of interlocus interactions in a complex disease: Rejection of the multiplicative model of epistasis in type 1 diabetes. *Genetics* **158,** 357.

Cornall, R. J., Prins, J. B., Todd, J. A., Pressey, A., DeLarato, N. H., Wicker, L. S., and Peterson, L. B. (1991). Type 1 diabetes in mice is linked to the interleukin-1 receptor and *Lsh/Ity/Bcg* genes on chromosome 1. *Nature* **353,** 262.

Cucca, F., Lampis, R., Congia, M., Angius, E., Nutland, S., Bain, S. C., Barnett, A. H., and Todd, J. A. (2001). A correlation between the relative predisposition of MHC class II alleles to type 1 diabetes and the structure of their proteins. *Hum. Mol. Genet.* **10,** 2025.

Denny, P., Lord, C. J., Hill, N. J., Goy, J. V., Levy, E. R., Podolin, P. L., Peterson, L. B., Wicker, L. S., Todd, J. A., and Lyons, P. A. (1997). Mapping of the IDDM locus *Idd3* to a 0.35-cM interval containing the interleukin-2 gene. *Diabetes* **46,** 695.

Deruytter, N., Boulard, O., and Garchon, H. J. (2004). Mapping non-class II *H2*-linked loci for type 1 diabetes in nonobese diabetic mice. *Diabetes* **53,** 3323.

Encinas, J. A., Wicker, L. S., Peterson, L. B., Mukasa, A., Teuscher, C., Sobel, R., Weiner, H. L., Seidman, C. E., Seidman, J. G., and Kuchroo, V. K. (1999). QTL influencing autoimmune diabetes and encephalomyelitis map to a 0.15-cM region containing *Il2*. *Nat. Genet.* **21,** 158.

Esposito, L., Hill, N. J., Pritchard, L. E., Cucca, F., Muxworthy, C., Merriman, M. E., Wilson, A., Julier, C., Delepine, M., Tuomilehto, J., Tuomilehto-Wolf, E., Ionesco-Tirgoviste, C., *et al.* (1998). Genetic analysis of chromosome 2 in type 1 diabetes: Analysis of putative loci *IDDM7*, *IDDM12*, and *IDDM13* and candidate genes *NRAMP1* and *IA-2* and the interleukin-1 gene cluster. IMDIAB Group. *Diabetes* **47,** 1797.

Fox, C. J., Hammerman, P. S., and Thompson, C. B. (2005). Fuel feeds function: Energy metabolism and the T-cell response. *Nat. Rev. Immunol.* **5,** 844.

Gavin, M. A., Rasmussen, J. P., Fontenot, J. D., Vasta, V., Manganiello, V. C., Beavo, J. A., and Rudensky, A. Y. (2007). Foxp3-dependent programme of regulatory T-cell differentiation. *Nature* **445,** 771.

Ghosh, S., Palmer, S. M., Rodrigues, N. R., Cordell, H. J., Hearne, C. M., Cornall, R. J., Prins, J. B., McShane, P., Lathrop, G. M., Peterson, L. B., Wicker, L. S., and Todd, J. A. (1993). Polygenic control of autoimmune diabetes in nonobese diabetic mice. *Nat. Genet.* **4,** 404.

Gonzalez, A., Katz, J. D., Mattei, M. G., Kikutani, H., Benoist, C., and Mathis, D. (1997). Genetic control of diabetes progression. *Immunity* **7,** 873.

Gregory, S. G., Schmidt, S., Seth, P., Oksenberg, J. R., Hart, J., Prokop, A., Caillier, S. J., Ban, M., Goris, A., Barcellos, L. F., Lincoln, R., McCauley, J. L., *et al.* (2007). Interleukin 7 receptor alpha chain (IL7R) shows allelic and functional association with multiple sclerosis. *Nat. Genet.* **39,** 1083.

Greve, B., Vijayakrishnan, L., Kubal, A., Sobel, R. A., Peterson, L. B., Wicker, L. S., and Kuchroo, V. K. (2004). The diabetes susceptibility locus *Idd5.1* on mouse chromosome 1 regulates ICOS expression and modulates murine experimental autoimmune encephalomyelitis. *J. Immunol.* **173,** 157.

Hamilton-Williams, E. E., Serreze, D. V., Charlton, B., Johnson, E. A., Marron, M. P., Mullbacher, A., and Slattery, R. M. (2001). Transgenic rescue implicates beta2-microglobulin as a diabetes susceptibility gene in nonobese diabetic (NOD) mice. *Proc. Natl. Acad. Sci. USA* **98,** 11533.

Hermann, R., Lipponen, K., Kiviniemi, M., Kakko, T., Veijola, R., Simell, O., Knip, M., and Ilonen, J. (2006). Lymphoid tyrosine phosphatase (LYP/PTPN22) Arg620Trp variant regulates insulin autoimmunity and progression to type 1 diabetes. *Diabetologia* **49,** 1198.

Hill, N. J., Lyons, P. A., Armitage, N., Todd, J. A., Wicker, L. S., and Peterson, L. B. (2000). NOD *Idd5* locus controls insulitis and diabetes and overlaps the orthologous *CTLA4/IDDM12* and *NRAMP1* loci in humans. *Diabetes* **49,** 1744.

Hollis-Moffatt, J. E., Hook, S. M., and Merriman, T. R. (2005). Colocalization of mouse autoimmune diabetes loci *Idd21.1* and *Idd21.2* with *IDDM6* (human) and *Iddm3* (rat). *Diabetes* **54,** 2820.

Howson, J. M., Dunger, D. B., Nutland, S., Stevens, H., Wicker, L. S., and Todd, J. A. (2007). A type 1 diabetes subgroup with a female bias is characterised by failure in tolerance to thyroid peroxidase at an early age and a strong association with the cytotoxic T-lymphocyte-associated antigen-4 gene. *Diabetologia* **50,** 741.

Hung, M. S., Avner, P., and Rogner, U. C. (2006). Identification of the transcription factor ARNTL2 as a candidate gene for the type 1 diabetes locus *Idd6*. *Hum. Mol. Genet.* **15,** 2732.

Hunter, K., Rainbow, D., Plagnol, V., Todd, J. A., Peterson, L. B., and Wicker, L. S. (2007). Interactions between *Idd5.1/Ctla4* and other type 1 diabetes genes. *J. Immunol.* **179,** 8341.

Irie, J., Wu, Y., Wicker, L. S., Rainbow, D., Nalesnik, M. A., Hirsch, R., Peterson, L. B., Leung, P. S., Cheng, C., Mackay, I. R., Gershwin, M. E., and Ridgway, W. M. (2006). NOD.c3c4 congenic mice develop autoimmune biliary disease that serologically and pathogenetically models human primary biliary cirrhosis. *J. Exp. Med.* **203,** 1209.

Irie, J., Reck, B., Wu, Y., Wicker, L. S., Howlett, S., Rainbow, D., Feingold, E., and Ridgway, W. M. (2008). Genome-wide microarray expression analysis of CD4+ T Cells from nonobese diabetic congenic mice identifies *Cd55* (*Daf1*) and *Acadl* as candidate genes for type 1 diabetes. *J. Immunol.* **180,** 1071.

Ivakine, E. A., Gulban, O. M., Mortin-Toth, S. M., Wankiewicz, E., Scott, C., Spurrell, D., Canty, A., and Danska, J. S. (2006). Molecular genetic analysis of the *Idd4* locus implicates the IFN response in type 1 diabetes susceptibility in nonobese diabetic mice. *J. Immunol.* **176,** 2976.

Jiang, W., Anderson, M. S., Bronson, R., Mathis, D., and Benoist, C. (2005). Modifier loci condition autoimmunity provoked by Aire deficiency. *J. Exp. Med.* **202,** 805.

Kissler, S., Stern, P., Takahashi, K., Hunter, K., Peterson, L. B., and Wicker, L. S. (2006). *In vivo* RNA interference demonstrates a role for Nramp1 in modifying susceptibility to type 1 diabetes. *Nat. Genet.* **38,** 479.

Koarada, S., Wu, Y., Fertig, N., Sass, D. A., Nalesnik, M., Todd, J. A., Lyons, P. A., Fenyk-Melody, J., Rainbow, D. B., Wicker, L. S., Peterson, L. B., and Ridgway, W. M. (2004). Genetic control of autoimmunity: Protection from diabetes, but spontaneous autoimmune biliary disease in a nonobese diabetic congenic strain. *J. Immunol.* **173,** 2315.

Liston, A., Lesage, S., Gray, D. H., O'Reilly, L. A., Strasser, A., Fahrer, A. M., Boyd, R. L., Wilson, J., Baxter, A. G., Gallo, E. M., Crabtree, G. R., Peng, K., *et al.* (2004). Generalized resistance to thymic deletion in the NOD mouse; a polygenic trait characterized by defective induction of Bim. *Immunity* **21,** 817.

Litherland, S. A., Grebe, K. M., Belkin, N. S., Paek, E., Elf, J., Atkinson, M., Morel, L., Clare-Salzler, M. J., and McDuffie, M. (2005). Nonobese diabetic mouse congenic analysis reveals chromosome 11 locus contributing to diabetes susceptibility, macrophage STAT5 dysfunction, and granulocyte-macrophage colony-stimulating factor overproduction. *J. Immunol.* **175,** 4561.

Lyons, P. A., and Wicker, L. S. (1999). Localising quantitative trait loci in the NOD mouse model of type 1 diabetes. *Curr. Dir. Autoimmun.* **1,** 208.

Lyons, P. A., Armitage, N., Argentina, F., Denny, P., Hill, N. J., Lord, C. J., Wilusz, M. B., Peterson, L. B., Wicker, L. S., and Todd, J. A. (2000a). Congenic mapping of the type 1 diabetes locus, *Idd3*, to a 780-kb region of mouse chromosome 3: Identification of a candidate segment of ancestral DNA by haplotype mapping. *Genome Res.* **10,** 446.

Lyons, P. A., Hancock, W. W., Denny, P., Lord, C. J., Hill, N. J., Armitage, N., Siegmund, T., Todd, J. A., Phillips, M. S., Hess, J. F., Chen, S. L., Fischer, P. A., *et al.* (2000b). The NOD *Idd9* genetic interval influences the pathogenicity of insulitis and contains molecular variants of *Cd30, Tnfr2*, and *Cd137*. *Immunity* **13,** 107.

Lyons, P. A., Armitage, N., Lord, C. J., Phillips, M. S., Todd, J. A., Peterson, L. B., and Wicker, L. S. (2001). Mapping by genetic interaction: High-resolution congenic mapping of the type 1 diabetes loci *Idd10* and *Idd18* in the NOD mouse. *Diabetes* **50,** 2633.

Manolio, T. A., Brooks, L. D., and Collins, F. S. (2008). A HapMap harvest of insights into the genetics of common disease. *J. Clin. Invest.* **118,** 1590.

Martinez, X., Kreuwel, H. T., Redmond, W. L., Trenney, R., Hunter, K., Rosen, H., Sarvetnick, N., Wicker, L. S., and Sherman, L. A. (2005). CD8+ T cell tolerance in nonobese diabetic mice is restored by insulin-dependent diabetes resistance alleles. *J. Immunol.* **175,** 1677.

Mathews, C. E., Graser, R. T., Bagley, R. J., Caldwell, J. W., Li, R., Churchill, G. A., Serreze, D. V., and Leiter, E. H. (2003). Genetic analysis of resistance to Type-1 diabetes in ALR/Lt mice, a NOD-related strain with defenses against autoimmune-mediated diabetogenic stress. *Immunogenetics* **55**, 491.

Matsubara, Y., Indo, Y., Naito, E., Ozasa, H., Glassberg, R., Vockley, J., Ikeda, Y., Kraus, J., and Tanaka, K. (1989). Molecular cloning and nucleotide sequence of cDNAs encoding the precursors of rat long chain acyl-coenzyme A, short chain acyl-coenzyme A, and isovaleryl-coenzyme A dehydrogenases. Sequence homology of four enzymes of the acyl-CoA dehydrogenase family. *J. Biol. Chem.* **264**, 16321.

McAleer, M. A., Reifsnyder, P., Palmer, S. M., Prochazka, M., Love, J. M., Copeman, J. B., Powell, E. E., Rodrigues, N. R., Prins, J. B., Serreze, D. V., Delarato, N. H., Wicker, L. S., Peterson, L. B., Schork, N. J., Todd, J. A., and Leiter, E. H. (1995). Crosses of NOD mice with the related NON strain. A polygenic model for IDDM. *Diabetes* **44**, 1186.

Morahan, G., McClive, P., Huang, D., Little, P., and Baxter, A. (1994). Genetic and physiological association of diabetes susceptibility with raised Na+/H+ exchange activity. *Proc. Natl. Acad. Sci. USA* **91**, 5898.

Morin, J., Boitard, C., Vallois, D., Avner, P., and Rogner, U. C. (2006). Mapping of the murine type 1 diabetes locus *Idd20* by genetic interaction. *Mamm. Genome* **17**, 1105.

Nejentsev, S., Howson, J. M., Walker, N. M., Szeszko, J., Field, S. F., Stevens, H. E., Reynolds, P., Hardy, M., King, E., Masters, J., Hulme, J., Maier, L. M., et al. (2007). Localization of type 1 diabetes susceptibility to the MHC class I genes *HLA-B* and *HLA-A*. *Nature* **450**, 887.

Pearce, R. B. (1998). Fine-mapping of the mouse T lymphocyte fraction (*Tlf*) locus on chromosome 9: Association with autoimmune diabetes. *Autoimmunity* **28**, 31.

Pearson, T., Weiser, P., Markees, T. G., Serreze, D. V., Wicker, L. S., Peterson, L. B., Cumisky, A. M., Shultz, L. D., Mordes, J. P., Rossini, A. A., and Greiner, D. L. (2004). Islet allograft survival induced by costimulation blockade in NOD mice is controlled by allelic variants of *Idd3*. *Diabetes* **53**, 1972.

Penha-Goncalves, C., Moule, C., Smink, L. J., Howson, J., Gregory, S., Rogers, J., Lyons, P. A., Suttie, J. J., Lord, C. J., Peterson, L. B., Todd, J. A., and Wicker, L. S. (2003). Identification of a structurally distinct CD101 molecule encoded in the 950-kb *Idd10* region of NOD mice. *Diabetes* **52**, 1551.

Podolin, P. L., Denny, P., Lord, C. J., Hill, N. J., Todd, J. A., Peterson, L. B., Wicker, L. S., and Lyons, P. A. (1997). Congenic mapping of the insulin-dependent diabetes (*Idd*) gene, *Idd10*, localizes two genes mediating the *Idd10* effect and eliminates the candidate *Fcgr1*. *J. Immunol.* **159**, 1835.

Podolin, P. L., Denny, P., Armitage, N., Lord, C. J., Hill, N. J., Levy, E. R., Peterson, L. B., Todd, J. A., Wicker, L. S., and Lyons, P. A. (1998). Localization of two insulin-dependent diabetes (*Idd*) genes to the *Idd10* region on mouse chromosome 3. *Mamm. Genome* **9**, 283.

Podolin, P. L., Wilusz, M. B., Cubbon, R. M., Pajvani, U., Lord, C. J., Todd, J. A., Peterson, L. B., Wicker, L. S., and Lyons, P. A. (2000). Differential glycosylation of interleukin 2, the molecular basis for the NOD *Idd3* type 1 diabetes gene? *Cytokine* **12**, 477.

Rainbow, D. B., Esposito, L., Howlett, S. K., Hunter, K. M., Todd, J. A., Peterson, L. B., and Wicker, L. S. (2008). Commonality in the genetic control of Type 1 diabetes in humans and NOD mice: Variants of genes in the IL-2 pathway are associated with autoimmune diabetes in both species. *Biochem. Soc. Trans.* **36**, 312.

Reifsnyder, P. C., Li, R., Silveira, P. A., Churchill, G., Serreze, D. V., and Leiter, E. H. (2005). Conditioning the genome identifies additional diabetes resistance loci in Type I diabetes resistant NOR/Lt mice. *Genes Immun.* **6**, 528.

Risch, N., Ghosh, S., and Todd, J. A. (1993). Statistical evaluation of multiple-locus linkage data in experimental species and its relevance to human studies: Application to nonobese diabetic (NOD) mouse and human insulin-dependent diabetes mellitus (IDDM). *Am. J. Hum. Genet.* **53**, 702.

Robles, D. T., Eisenbarth, G. S., Dailey, N. J., Peterson, L. B., and Wicker, L. S. (2003). Insulin autoantibodies are associated with islet inflammation but not always related to diabetes progression in NOD congenic mice. *Diabetes* **52,** 882.

Rogner, U. C., Boitard, C., Morin, J., Melanitou, E., and Avner, P. (2001). Three loci on mouse chromosome 6 influence onset and final incidence of type I diabetes in NOD.C3H congenic strains. *Genomics* **74,** 163.

Runstadler, J. A., Saila, H., Savolainen, A., Leirisalo-Repo, M., Aho, K., Tuomilehto-Wolf, E., Tuomilehto, J., and Seldin, M. F. (2005). Association of SLC11A1 (NRAMP1) with persistent oligoarticular and polyarticular rheumatoid factor-negative juvenile idiopathic arthritis in Finnish patients: Haplotype analysis in Finnish families. *Arthritis Rheum.* **52,** 247.

Serreze, D. V., Bridgett, M., Chapman, H. D., Chen, E., Richard, S. D., and Leiter, E. H. (1998). Subcongenic analysis of the *Idd13* locus in NOD/Lt mice: Evidence for several susceptibility genes including a possible diabetogenic role for beta 2-microglobulin. *J. Immunol.* **160,** 1472.

Serreze, D. V., Choisy-Rossi, C. M., Grier, A. E., Holl, T. M., Chapman, H. D., Gahagan, J. R., Osborne, M. A., Zhang, W., King, B. L., Brown, A., Roopenian, D., and Marron, M. P. (2008). Through regulation of TCR expression levels, an *Idd7* region gene(s) interactively contributes to the impaired thymic deletion of autoreactive diabetogenic CD8+ T cells in nonobese diabetic mice. *J. Immunol.* **180,** 3250.

Shaw, M. A., Clayton, D., Atkinson, S. E., Williams, H., Miller, N., Sibthorpe, D., and Blackwell, J. M. (1996). Linkage of rheumatoid arthritis to the candidate gene *NRAMP1* on 2q35. *J. Med. Genet.* **33,** 672.

Siegmund, T., Armitage, N., Wicker, L. S., Peterson, L. B., Todd, J. A., and Lyons, P. A. (2000). Analysis of the mouse CD30 gene: A candidate for the NOD mouse type 1 diabetes locus *Idd9.2*. *Diabetes* **49,** 1612.

Silveira, P. A., Chapman, H. D., Stolp, J., Johnson, E., Cox, S. L., Hunter, K., Wicker, L. S., and Serreze, D. V. (2006). Genes within the *Idd5* and *Idd9/11* diabetes susceptibility loci affect the pathogenic activity of B cells in nonobese diabetic mice. *J. Immunol.* **177,** 7033.

Singer, S. M., Tisch, R., Yang, X. D., and McDevitt, H. O. (1993). An $A\beta^d$ transgene prevents diabetes in nonobese diabetic mice by inducing regulatory T cells. *Proc. Natl. Acad. Sci. USA* **90,** 9566.

Slattery, R. M., Kjer-Nielsen, L., Allison, J., Charlton, B., Mandel, T. E., and Miller, J. F. (1990). Prevention of diabetes in non-obese diabetic I-A^k transgenic mice. *Nature* **345,** 724.

Smyth, D. J., Cooper, J. D., Howson, J. M., Walker, N. M., Plagnol, V., Stevens, H., Clayton, D. G., and Todd, J. A. (2008). PTPN22 Trp620 explains the association of chromosome 1p13 with type 1 diabetes and shows a statistical interaction with HLA class II genotypes. *Diabetes* **57,** 1730.

Steck, A. K., Liu, S. Y., McFann, K., Barriga, K. J., Babu, S. R., Eisenbarth, G. S., Rewers, M. J., and She, J. X. (2006). Association of the PTPN22/LYP gene with type 1 diabetes. *Pediatr. Diabetes* **7,** 274.

Stober, C. B., Brode, S., White, J. K., Popoff, J. F., and Blackwell, J. M. (2007). Slc11a1, formerly Nramp1, is expressed in dendritic cells and influences major histocompatibility complex class II expression and antigen-presenting cell function. *Infect. Immun.* **75,** 5059.

Teuscher, C., Wardell, B. B., Lunceford, J. K., Michael, S. D., and Tung, K. S. (1996). *Aod2*, the locus controlling development of atrophy in neonatal thymectomy-induced autoimmune ovarian dysgenesis, co-localizes with *Il2*, *Fgfb*, and *Idd3*. *J. Exp. Med.* **183,** 631.

Todd, J. A., Aitman, T. J., Cornall, R. J., Ghosh, S., Hall, J. R., Hearne, C. M., Knight, A. M., Love, J. M., McAleer, M. A., Prins, J. B., Rodrigues, N., Lathrop, M., Pressey, A., Delarato, N. H., Peterson, L. B., and Wicker, L. S. (1991). Genetic analysis of autoimmune type 1 diabetes mellitus in mice. *Nature* **351,** 542.

Todd, J. A., Walker, N. M., Cooper, J. D., Smyth, D. J., Downes, K., Plagnol, V., Bailey, R., Nejentsev, S., Field, S. F., Payne, F., Lowe, C. E., Szeszko, J. S., *et al.* (2007). Robust associations of four new chromosome regions from genome-wide analyses of type 1 diabetes. *Nat. Genet.* **39,** 857.

Ueda, H., Howson, J. M., Esposito, L., Heward, J., Snook, H., Chamberlain, G., Rainbow, D. B., Hunter, K. M., Smith, A. N., Di Genova, G., Herr, M. H., Dahlman, I., et al. (2003). Association of the T-cell regulatory gene CTLA4 with susceptibility to autoimmune disease. Nature **423**, 506.

Vijayakrishnan, L., Slavik, J. M., Illes, Z., Greenwald, R. J., Rainbow, D., Greve, B., Peterson, L. B., Hafler, D. A., Freeman, G. J., Sharpe, A. H., Wicker, L. S., and Kuchroo, V. K. (2004). An autoimmune disease-associated CTLA-4 splice variant lacking the B7 binding domain signals negatively in T cells. Immunity **20**, 563.

Wellcome Trust Case Control Consortium (2007). Genome-wide association study of 14,000 cases of seven common diseases and 3,000 shared controls. Nature **447**, 661.

Wicker, L. S., Miller, B. J., Coker, L. Z., McNally, S. E., Scott, S., Mullen, Y., and Appel, M. C. (1987). Genetic control of diabetes and insulitis in the nonobese diabetic (NOD) mouse. J. Exp. Med. **165**, 1639.

Wicker, L. S., Miller, B. J., Fischer, P. A., Pressey, A., and Peterson, L. B. (1989). Genetic control of diabetes and insulitis in the nonobese diabetic mouse. Pedigree analysis of a diabetic H-$2^{nod/b}$ heterozygote. J. Immunol. **142**, 781.

Wicker, L. S., Todd, J. A., Prins, J. B., Podolin, P. L., Renjilian, R. J., and Peterson, L. B. (1994). Resistance alleles at two non-major histocompatibility complex-linked insulin-dependent diabetes loci on chromosome 3, Idd3 and Idd10, protect nonobese diabetic mice from diabetes. J. Exp. Med. **180**, 1705.

Wicker, L. S., Todd, J. A., and Peterson, L. B. (1995). Genetic control of autoimmune diabetes in the NOD mouse. Annu. Rev. Immunol. **13**, 179.

Wicker, L. S., Chamberlain, G., Hunter, K., Rainbow, D., Howlett, S., Tiffen, P., Clark, J., Gonzalez-Munoz, A., Cumiskey, A. M., Rosa, R. L., Howson, J. M., Smink, L. J., et al. (2004). Fine mapping, gene content, comparative sequencing, and expression analyses support Ctla4 and Nramp1 as candidates for Idd5.1 and Idd5.2 in the nonobese diabetic mouse. J. Immunol. **173**, 164.

Wicker, L. S., Clark, J., Fraser, H. I., Garner, V. E., Gonzalez-Munoz, A., Healy, B., Howlett, S., Hunter, K., Rainbow, D., Rosa, R. L., Smink, L. J., Todd, J. A., et al. (2005). Type 1 diabetes genes and pathways shared by humans and NOD mice. J. Autoimmun. **25**(Suppl), 29.

Yamanouchi, J., Rainbow, D., Serra, P., Howlett, S., Hunter, K., Garner, V. E., Gonzalez-Munoz, A., Clark, J., Veijola, R., Cubbon, R., Chen, S. L., Rosa, R., et al. (2007). Interleukin-2 gene variation impairs regulatory T cell function and causes autoimmunity. Nat. Genet. **39**, 329.

Yui, M. A., Sharp, L. L., Havran, W. L., and Rothenberg, E. V. (2004). Preferential activation of an IL-2 regulatory sequence transgene in TCR gamma delta and NKT cells: Subset-specific differences in IL-2 regulation. J. Immunol. **172**, 4691.

Zheng, Z. M. (2004). Regulation of alternative RNA splicing by exon definition and exon sequences in viral and mammalian gene expression. J. Biomed. Sci. **11**, 278.

Zucchelli, S., Holler, P., Yamagata, T., Roy, M., Benoist, C., and Mathis, D. (2005). Defective central tolerance induction in NOD mice: Genomics and genetics. Immunity **22**, 385.

SUBJECT INDEX

A

Activation induced cell death (AICD)
Antigen presenting cells (APCs)
  driving epitope spread, key roles, 127
  pMHC complexes, 130
Antinuclear autoantibodies (ANA), 102
Autoimmune diabetes mellitus
  diabetic antigens and diabetogenic
      T cells, 3
    autoimmune repertoire, 4
    CD8 T cells, 6
    β cell antigens, presentation, 4
    central thymic problem, 5
    insulin reactive, 5
  histocompatibility molecules, 3
  type 1 diabetes, genetics of, 6–7
Autoimmune disease, 126
Autoimmune polyendocrinopathy-
    candidiasis-ectodermal dystrophy
    (APECED), 128
Autoimmune regulatory (AIRE) gene,
    128–129, 152

B

Bio breeding (BB), 14

C

CD3 monoclonal antibody therapy, 29
CD8+ T cells
  CD4+ T cells, requirement
    CD40/CD154 interactions, 100
    diabetes retardation, 99
  cross-tolerance
    CTLA-4, hyper-lymphoproliferation,
        103
    PD-1/PD-L1 interaction, 103–104
  defective immune tolerance
    α-GalCer-mediated protection, 102–103
    T cell-intrinsic defects, 102
  diabetogenic
    autoantigenic pMHC complexes, 98–99
    in PLNs, 99

  immunologic tolerance
    autoreactive T cells, anergy, 108
    diabetes protection, 111
    $IGRP_{206-214}$ mimotope, administration,
        109
    peptide ligands, 108–109
    selective expansion, 110
  inflammatory signals
    endogenous adjuvants, 101
    immune-complex antibodies,
        100–101
  recruitment to islets
    chemokines production, 104
    islet homing, 105
  TCR-transgenic models
    diabetes and, 97
    neo-self antigens, 98
    TCRα and TCRβ rearrangements, 97
Chronic insulin therapy, 14
Cytotoxic T lymphocyte (CTL), 83

D

Dendritic cells (DC)
  immunoregulation
    nonautoimmune mice, 139
    protein tyrosine phosphatase, 140
    transcription factor, 138
  and macrophages, 127
  thymocyte negative selection, 128–129
Diabetes autoantibody standardization
    program (DASP), 58
Diabetes prevention trial (DPT1), 15
Dystrophia myotonica kinase
    ($DMK_{138-146}$), 98

E

ELISpot-based approach, CD8+ T cell
    assay, 86
Enzyme-linked immunosorbent assay
    (ELISA)
  disadvantages of, 58
  GAD 65 antibody detection, 57
Enzyme-linked immunospot (ELISPOT), 55

Epstein Barr virus (EBV)
  transient reactivation, 17–18
  viral re-activation, 28
Exonic splicing enhancer (ESE), 161
Exonic splicing silencer (ESS), 161
Experimental allergic encephalomyelitis (EAE), 20
Experimental autoimmune encephalomyelitis, 164

G

GAD 65 cDNA, 43–44
Genome wide association studies (GWAS), 152
Glial fibrillary acidic protein (GFAP), 85, 87, 88, 90, 109
  $H-2K^d$-restricted epitopes, 91–92
  peri-islet Schwann cells (pSC), 91
Glutamic acid decarboxylase 65 (GAD 65), 8, 127–128, 136
  anti-idiotypic antibodies in T1D, 61
  autoantibodies and T1D, 59
  GAD 65A
    antibody epitopes, 59
    WHO standard, 58–59
  and GAD 67 gene structure and function
    crystal, 46–48
    genomic, 43–44
    primary, 44–46
  human monoclonal antibodies
    IgM and IgG, 59–60
    reactivity, 60–61
  NOD diabetes, 64
Glycosuria, 7–9

H

Heat shock proteins (HSPs), 101

I

I-Ag^7, class II MHC molecule, 3
Immunology of Diabetes Society (IDS), 56, 58
Insulin-dependent diabetes (IDD)
  IDD3
    cellular mechanisms, 165
    NOD congenic mice, functional studies, 162
    protective alleles, 163
    in vivo regulation, 164
  IDD5
    human genetic studies, 169
    immune system, 165
    protective alleles, 163, 168
    subregions, 168–169
  IDD5.3 and IDD5.4
    congenic strains, 168
    genetic complexity, masking and discovery, 167
  regions, 158–160
Insulinoma-associated antigen-2 (IA-2), 57, 92
Invariant NKT cells (iNKT cells), 137
Islet-associated glucose-6- phosphatase catalytic subunit-related protein (IGRP), 84, 92–93
Islet cell antibodies (ICA), 41
Islet cell surface antibodies (ICSA), 41

K

65 kDa glutamic acid decarboxylase (GAD 65), 87
64K protein, 40

L

Latent autoimmune diabetes in the adult (LADA), 40
Ligand-independent CTLA-4 (liCTLA-4), 131
Lymphocytic choriomeningitis virus glycoprotein (LCMV-GP), 98, 101, 104, 108
Lymphocytic choriomeningitis virus nucleoprotein (LCMV-NP), 98

M

Major histocompatibility (MHC)
Mixed meal tolerance test, 67
Monoclonal islet cell antibodies (MICA), 60
Myelin oligodendrocyte glycoprotein (MOG), 20

N

Nonobese diabetic (NOD), 14
  BioBreeder (BB) rat and, 2
  congenic mouse, 156
  models, 153–155
  mouse, 81
  rodent models, 126
  therapies development
    C-peptide levels, 8
    monoclonal antibody administration, 7

Subject Index

P

Peptide/major histocompatibility complex (pMHC) molecules, 80
Peripheral blood mononuclear cells (PBMCs), 86
Polymerase chain reaction (PCR), 17
Programmed death 1(PD-1), 103–104

R

Radiobinding assays (RBA)
    GAD65 antibody detection, 57
    and protein A, 59
Rat insulin promoter (RIP), 98
Regulatory T cells (Tregs), 81–82, 103, 108

S

Severe-combined immunodeficient NOD (NOD.scid), 83
Single nucleotide polymorphisms (SNPs), 152, 164
Streptozotocin drug, inflammatory response, 2
Subcutaneous immunotherapy (SCIT), 68

T

T cell receptor (TCR), 49
    and CD28, 131
    negative regulator, signaling events, 129
    self-peptide/MHC molecules (self-pMHC), 128
    transgenic mice, 130
T-cells
    immunoregulation
        $CD4^+$ and $CD8^+$, 139
        conventional Treg, 136–138
        $Foxp3^+$ Treg, 132–136
        lymphopenia, 135
        Th1 and Th2 cells, 132
        in vitro and *in vivo* conditions, 133–134
    in vitro, 131, 135
Thymic epithelial cells (TEC)
Time-resolved fluorescence (TRF)
    advantages and disadvantages, 58
    GAD65 antibody detection, 57
Type 1 diabetes mellitus (T1DM)
    autoimmune process
        β cell autoimmunity, 126
        intra- and inter-molecular epitopes, 127
        T cells functions, 128
    central T cell tolerance, dysregulation
        apoptotic inducing events, 128
        LYP functions in, 129
    genetics of
        C-peptide levels, 8
        genes identification, 6
        in NOD mouse and human, 7
    glutamic acid decarboxylase 65 (GAD 65), administration
        dosage, 8
        purified form, 8
    insulin producing β cells, 126
    models of
        animal, 2
        BioBreeder(BB) rat, 2–3
    NOD mouse model
        affecting frequency, 157
        alleles altering functions, 161–162
        anti-thyroid autoantibodies, 169
        autoimmune diseases, 152
        B cells, *Idd* regions, 162
        biological pathways, 163
        CD4 and CD8 T cells, 161
        cell transfer experiments, 156
        congenic mapping and intervals, 162
        cytotoxic effector cells, 165
        disease-resistant *vs.* disease-susceptible alleles, 164
        genetic control, 155
        human disease-causing genes, 153
        IDD3, 162–165
        IDD5, 162–163, 165–167
        IDD5.3 and IDD5.4 masking, 167–168
        IL2/IDD3, IDD5 subregions, 168–169
        large-scale linkage study, 154
        multigenic disease, 153–154
        non-MHC susceptibility alleles, 170
        protective *vs.* susceptibility alleles, 168
        Salmonella infection, 167
        TcR zeta chain, 166
    peripheral T cell tolerance, dysregulation
        anergy and deletion, 130–132
        immunoregulation defects, 132–140
        naive T cells, signals, 130
        negative regulatory receptors, 131
        PD-L1 blocks, 132
    prevention, humans
        autoantigens, 10
        β cell destruction, 9
        HLA genotypes, 9
Type 1 diabetes mellitus (T1DM), CD3 antibodies
    antigen specific and long lasting
        cellular and molecular mechanism, 23
        cyclosporin and cyclophosphamide in, 26
        FcR nonbinding antibodies, 21–22

Type 1 diabetes mellitus (T1DM), CD3 antibodies (cont.)
  immune privilege, 25–26
  PDL-1 and ICOS ligand expression, 24–25
  peripheral tolerance mechanisms, 22–23
  TCR/CD3 receptor, antigenic modulation, 22
  TGF-β, 23–24
  tolerogenic capacity, 21
  tolerogenic effect, 23, 24, 27
  tregs in, 25
ChAglyCD3, 16–17
chronic insulin therapy, 14–15
clinical results
  Fc-mutated monoclonal antibodies, 20–21
  heart allografts, 19
  syngeneic islet grafts, 20
  transplantation, 18–19
hOKT3g1, 16
immunotherapy, 15
prediabetes, 14
side effects, 17
therapy
  autoimmunity recurrence, 29–30
  B cells, combined targeting, 30
  EBV viral re-activation, 28
  immunization, 28–29
  long-term effect of, 17
  T-cell activation, 26
  therapeutic activity, 29
Type 1 diabetes mellitus (T1DM), CD8+ T cells
  autoreactive
    diabetic patients, 83
    IGRP$_{206-214}$-reactive, 84, 86
    insulin derived epitope, 84
    NOD neonates, 83
  β-cell cytotoxicity, mechanisms of
    Fas vs granule exocytosis, 105–106
    TNF-α, TNFR, IFN-γ, and IL-1, 107–108
  β-cell-specific, relative contribution
    proinsulin and IGRP antigens, 95
    T cell clones side, 96
  diabetogenic antigens, in humans
    GAD65, 91
    GFAP, 91–92
    HLA-A2-restricted epitopes, identification, 86
    IGRP, 92–93
    insulin, 93–94
    Insulinoma-associated antigen-2 (IA-2), 92
  MHC-stabilization assay, 86–87
  peptide epitopes, 87
  H-2K^d and H-2D^b-restricted epitopes, 85
  HLA class I-restricted epitopes, 88–90
  MHC class I and
    antigenic peptide presentation, 82–83
    CD4+ T cells, 82
    polyclonal autoimmune response, 81
  negative selection, 80
Type 1 diabetes mellitus (T1DM), GAD65 autoimmunity
  cellular
    antigen presenting cells, role, 48–52
    antigen-specific regulatory T cells, phenotyping, 54–55
    antigen-specific T-cell phenotyping, 53–54
    enzyme-linked immunospot (ELISPOT), 55
    immunology of diabetes (IDS), T-cell, 56
    T and B cell interactions, 55–56
    T-cell proliferation, 52
  clinical studies, ongoing and future, 67
  and GAD67 structure and function, 43–48
  humoral
    anti-idiotypic antibodies, 61
    autoantibodies and, 59
    autoantibody assays, 56–58
    human monoclonal antibodies, 59–61
    IDW workshops and DASP, 58–59
  64K protein
    GAD activity in, 42–43
    islet autoantibodies description, 41
  pathogenesis
    effector and regulatory characteristics, 62–63
    (pro)insulin autoantibodies, 61
    T-cell level cross-reactivity, 61–62
  phase I clinical studies
    alum-formulated recombinant human GAD65, use, 64
    rhGAD65 Bulk Product, 65
  phase II clinical studies
    alum-formulated recombinant human GAD65, 65–66
    Fasting c-peptide levels, 66
    GAD65Aepitope recognition, 67
  preclinical studies, 64

V

Variable nucleotide tandem repeat (VNTR)

Contents of Recent Volumes

Volume 85

Cumulative Subject Index Volumes 66–82

Volume 86

Adenosine Deaminase Deficiency: Metabolic Basis of Immune Deficiency and Pulmonary Inflammation
Michael R. Blackburn and Rodney E. Kellems

Mechanism and Control of V(D)J Recombination Versus Class Switch Recombination: Similarities and Differences
Darryll D. Dudley, Jayanta Chaudhuri, Craig H. Bassing, and Frederick W. Alt

Isoforms of Terminal Deoxynucleotidyltransferase: Developmental Aspects and Function
To-Ha Thai and John F. Kearney

Innate Autoimmunity
Michael C. Carroll and V. Michael Holers

Formation of Bradykinin: A Major Contributor to the Innate Inflammatory Response
Kusumam Joseph and Allen P. Kaplan

Interleukin-2, Interleukin-15, and Their Roles in Human Natural Killer Cells
Brian Becknell and Michael A. Caligiuri

Regulation of Antigen Presentation and Cross-Presentation in the Dendritic Cell Network: Facts, Hypothesis, and Immunological Implications
Nicholas S. Wilson and Jose A. Villadangos

Index

Volume 87

Role of the LAT Adaptor in T-Cell Development and T_h2 Differentiation
Bernard Malissen, Enrique Aguado, and Marie Malissen

The Integration of Conventional and Unconventional T Cells that Characterizes Cell-Mediated Responses
Daniel J. Pennington, David Vermijlen, Emma L. Wise, Sarah L. Clarke, Robert E. Tigelaar, and Adrian C. Hayday

Negative Regulation of Cytokine and TLR Signalings by SOCS and Others
Tetsuji Naka, Minoru Fujimoto, Hiroko Tsutsui, and Akihiko Yoshimura

Pathogenic T-Cell Clones in Autoimmune Diabetes: More Lessons from the NOD Mouse
Kathryn Haskins

The Biology of Human Lymphoid Malignancies Revealed by Gene Expression Profiling
Louis M. Staudt and Sandeep Dave

New Insights into Alternative Mechanisms of Immune Receptor Diversification
Gary W. Litman, John P. Cannon, and Jonathan P. Rast

The Repair of DNA Damages/ Modifications During the Maturation of the Immune System: Lessons from Human Primary Immunodeficiency Disorders and Animal Models
Patrick Revy, Dietke Buck, Françoise le Deist, and Jean-Pierre de Villartay

Antibody Class Switch Recombination: Roles for Switch Sequences and Mismatch Repair Proteins
Irene M. Min and Erik Selsing

Index

Volume 88

CD22: A Multifunctional Receptor That Regulates B Lymphocyte Survival and Signal Transduction
Thomas F. Tedder, Jonathan C. Poe, and Karen M. Haas

Tetramer Analysis of Human Autoreactive CD4-Positive T Cells
Gerald T. Nepom

Regulation of Phospholipase C-γ2 Networks in B Lymphocytes
Masaki Hikida and Tomohiro Kurosaki

Role of Human Mast Cells and Basophils in Bronchial Asthma
Gianni Marone, Massimo Triggiani, Arturo Genovese, and Amato De Paulis

A Novel Recognition System for MHC Class I Molecules Constituted by PIR
Toshiyuki Takai

Dendritic Cell Biology
Francesca Granucci, Maria Foti, and Paola Ricciardi-Castagnoli

The Murine Diabetogenic Class II Histocompatibility Molecule I-A^{g7}: Structural and Functional Properties and Specificity of Peptide Selection
Anish Suri and Emil R. Unanue

RNAi and RNA-Based Regulation of Immune System Function
Dipanjan Chowdhury and Carl D. Novina

Index

Volume 89

Posttranscriptional Mechanisms Regulating the Inflammatory Response
Georg Stoecklin Paul Anderson

Negative Signaling in Fc Receptor Complexes
Marc Daëron and Renaud Lesourne

The Surprising Diversity of Lipid Antigens for CD1-Restricted T Cells
D. Branch Moody

Lysophospholipids as Mediators of Immunity
Debby A. Lin and Joshua A. Boyce

Systemic Mastocytosis
Jamie Robyn and Dean D. Metcalfe

Regulation of Fibrosis by the Immune System
Mark L. Lupher, Jr. and W. Michael Gallatin

Immunity and Acquired Alterations in Cognition and Emotion: Lessons from SLE
Betty Diamond, Czeslawa Kowal, Patricio T. Huerta, Cynthia Aranow, Meggan Mackay, Lorraine A. DeGiorgio, Ji Lee, Antigone Triantafyllopoulou, Joel Cohen-Solal Bruce, and T. Volpe

Immunodeficiencies with Autoimmune Consequences
Luigi D. Notarangelo, Eleonora Gambineri, and Raffaele Badolato

Index

Volume 90

Cancer Immunosurveillance and Immunoediting: The Roles of Immunity in Suppressing Tumor Development and Shaping Tumor Immunogenicity
Mark J. Smyth, Gavin P. Dunn, and Robert D. Schreiber

Mechanisms of Immune Evasion by Tumors
Charles G. Drake, Elizabeth Jaffee, and Drew M. Pardoll

Development of Antibodies and Chimeric Molecules for Cancer Immunotherapy
Thomas A. Waldmann and John C. Morris

Induction of Tumor Immunity Following Allogeneic Stem Cell Transplantation
Catherine J. Wu and Jerome Ritz

Vaccination for Treatment and Prevention of Cancer in Animal Models

Federica Cavallo, Rienk Offringa,
Sjoerd H. van der Burg, Guido Forni,
and Cornelis J. M. Melief

Unraveling the Complex Relationship Between Cancer Immunity and Autoimmunity: Lessons from Melanoma and Vitiligo
Hiroshi Uchi, Rodica Stan, Mary Jo Turk, Manuel E. Engelhorn, Gabrielle A. Rizzuto, Stacie M. Goldberg, Jedd D. Wolchok, and Alan N. Houghton

Immunity to Melanoma Antigens: From Self-Tolerance to Immunotherapy
Craig L. Slingluff, Jr., Kimberly A. Chianese-Bullock, Timothy N. J. Bullock, William W. Grosh, David W. Mullins, Lisa Nichols, Walter Olson, Gina Petroni, Mark Smolkin, and Victor H. Engelhard

Checkpoint Blockade in Cancer Immunotherapy
Alan J. Korman, Karl S. Peggs, and James P. Allison

Combinatorial Cancer Immunotherapy
F. Stephen Hodi and Glenn Dranoff

Index

Volume 91

A Reappraisal of Humoral Immunity Based on Mechanisms of Antibody-Mediated Protection Against Intracellular Pathogens
Arturo Casadevall and Liise-anne Pirofski

Accessibility Control of V(D)J Recombination
Robin Milley Cobb, Kenneth J. Oestreich, Oleg A. Osipovich, and Eugene M. Oltz

Targeting Integrin Structure and Function in Disease
Donald E. Staunton, Mark L. Lupher, Robert Liddington, and W. Michael Gallatin

Endogenous TLR Ligands and Autoimmunity
Hermann Wagner

Genetic Analysis of Innate Immunity
Kasper Hoebe, Zhengfan Jiang, Koichi Tabeta, Xin Du, Philippe Georgel, Karine Crozat, and Bruce Beutler

TIM Family of Genes in Immunity and Tolerance
Vijay K. Kuchroo, Jennifer Hartt Meyers, Dale T. Umetsu, and Rosemarie H. DeKruyff

Inhibition of Inflammatory Responses by Leukocyte Ig-Like Receptors
Howard R. Katz

Index

Volume 92

Systemic Lupus Erythematosus: Multiple Immunological Phenotypes in a Complex Genetic Disease
Anna-Marie Fairhurst, Amy E. Wandstrat, and Edward K. Wakeland

Avian Models with Spontaneous Autoimmune Diseases
Georg Wick, Leif Andersson, Karel Hala, M. Eric Gershwin, Carlo Selmi, Gisela F. Erf, Susan J. Lamont, and Roswitha Sgonc

Functional Dynamics of Naturally Occurring Regulatory T Cells in Health and Autoimmunity
Megan K. Levings, Sarah Allan, Eva d'Hennezel, and Ciriaco A. Piccirillo

BTLA and HVEM Cross Talk Regulates Inhibition and Costimulation

Maya Gavrieli, John Sedy, Christopher A. Nelson, and Kenneth M. Murphy

The Human T Cell Response to Melanoma Antigens
Pedro Romero, Jean-Charles Cerottini, and Daniel E. Speiser

Antigen Presentation and the Ubiquitin-Proteasome System in Host–Pathogen Interactions
Joana Loureiro and Hidde L. Ploegh

Index

Volume 93

Class Switch Recombination: A Comparison Between Mouse and Human
Qiang Pan-Hammarström, Yaofeng Zhao, and Lennart Hammarström

Anti-IgE Antibodies for the Treatment of IgE-Mediated Allergic Diseases
Tse Wen Chang, Pheidias C. Wu, C. Long Hsu, and Alfur F. Hung

Immune Semaphorins: Increasing Members and Their Diverse Roles
Hitoshi Kikutani, Kazuhiro Suzuki, and Atsushi Kumanogoh

Tec Kinases in T Cell and Mast Cell Signaling
Martin Felices, Markus Falk, Yoko Kosaka, and Leslie J. Berg

Integrin Regulation of Lymphocyte Trafficking: Lessons from Structural and Signaling Studies
Tatsuo Kinashi

Regulation of Immune Responses and Hematopoiesis by the Rap1 Signal
Nagahiro Minato, Kohei Kometani, and Masakazu Hattori

Lung Dendritic Cell Migration
Hamida Hammad and Bart N. Lambrecht

Index

Volume 94

Discovery of Activation-Induced Cytidine Deaminase, the Engraver of Antibody Memory
Masamichi Muramatsu, Hitoshi Nagaoka, Reiko Shinkura, Nasim A. Begum, and Tasuku Honjo

DNA Deamination in Immunity: AID in the Context of Its APOBEC Relatives
Silvestro G. Conticello, Marc-Andre Langlois, Zizhen Yang, and Michael S. Neuberger

The Role of Activation-Induced Deaminase in Antibody Diversification and Chromosome Translocations
Almudena Ramiro, Bernardo Reina San-Martin, Kevin McBride, Mila Jankovic, Vasco Barreto, André Nussenzweig, and Michel C. Nussenzweig

Targeting of AID-Mediated Sequence Diversification by cis-Acting Determinants
Shu Yuan Yang and David G. Schatz

AID-Initiated Purposeful Mutations in Immunoglobulin Genes
Myron F. Goodman, Matthew D. Scharff, and Floyd E. Romesberg

Evolution of the Immunoglobulin Heavy Chain Class Switch Recombination Mechanism
Jayanta Chaudhuri, Uttiya Basu, Ali Zarrin, Catherine Yan, Sonia Franco, Thomas Perlot, Bao Vuong, Jing Wang, Ryan T. Phan, Abhishek Datta, John Manis, and Frederick W. Alt

Beyond SHM and CSR: AID and Related Cytidine Deaminases in the Host Response to Viral Infection
Brad R. Rosenberg and F. Nina Papavasiliou

Role of AID in Tumorigenesis
Il-mi Okazaki, Ai Kotani, and Tasuku Honjo

Pathophysiology of B-Cell Intrinsic
  Immunoglobulin Class Switch
  Recombination Deficiencies
*Anne Durandy, Nadine Taubenheim,
  Sophie Peron, and Alain Fischer*

Index

Volume 95

Fate Decisions Regulating Bone Marrow
  and Peripheral B Lymphocyte
  Development
John G. Monroe and Kenneth Dorshkind

Tolerance and Autoimmunity:
  Lessons at the Bedside of Primary
  Immunodeficiencies
*Magda Carneiro-Sampaio and Antonio
  Coutinho*

B-Cell Self-Tolerance in Humans
*Hedda Wardemann and Michel
  C. Nussenzweig*

Manipulation of Regulatory T-Cell
  Number and Function with CD28-
  Specific Monoclonal Antibodies
Thomas Hünig

Osteoimmunology: A View from the Bone
Jean-Pierre David

Mast Cell Proteases
*Gunnar Pejler, Magnus Åbrink,
  Maria Ringvall, and Sara Wernersson*

Index

Volume 96

New Insights into Adaptive Immunity
  in Chronic Neuroinflammation
*Volker Siffrin, Alexander U. Brandt,
  Josephine Herz, and Frauke Zipp*

Regulation of Interferon-γ During Innate
  and Adaptive Immune Responses
*Jamie R. Schoenborn and Christopher
  B. Wilson*

The Expansion and Maintenance of
  Antigen-Selected CD8$^+$ T Cell Clones
Douglas T. Fearon

Inherited Complement Regulatory
  Protein Deficiency Predisposes to
  Human Disease in Acute Injury and
  Chronic Inflammatory States
*Anna Richards, David Kavanagh,
  and John P. Atkinson*

Fc-Receptors as Regulators of Immunity
Falk Nimmerjahn and Jeffrey V. Ravetch

Index

Volume 97

T Cell Activation and the Cytoskeleton:
  You Can't Have One Without
  the Other
Timothy S. Gomez and Daniel D. Billadeau

HLA Class II Transgenic Mice Mimic
  Human Inflammatory Diseases
*Ashutosh K. Mangalam, Govindarajan
  Rajagopalan, Veena Taneja, and
  Chella S. David*

Roles of Zinc and Zinc Signaling in
  Immunity: Zinc as an Intracellular
  Signaling Molecule
*Toshio Hirano, Masaaki Murakami,
  Toshiyuki Fukada, Keigo Nishida,
  Satoru Yamasaki, and Tomoyuki Suzuki*

The SLAM and SAP Gene Families
  Control Innate and Adaptive
  Immune Responses
*Silvia Calpe, Ninghai Wang,
  Xavier Romero, Scott B. Berger,
  Arpad Lanyi, Pablo Engel, and
  Cox Terhorst*

Conformational Plasticity and
  Navigation of Signaling Proteins
  in Antigen-Activated B Lymphocytes
*Niklas Engels, Michael Engelke, and
  Jürgen Wienands*

Index

Volume 98

Immune Regulation by B Cells and Antibodies: A View Towards the Clinic
Kai Hoehlig, Vicky Lampropoulou, Toralf Roch, Patricia Neves, Elisabeth Calderon-Gomez, Stephen M. Anderton, Ulrich Steinhoff, and Simon Fillatreau

Cumulative Environmental Changes, Skewed Antigen Exposure, and the Increase of Allergy
Tse Wen Chang and Ariel Y. Pan

New Insights on Mast Cell Activation via the High Affinity Receptor for IgE
Juan Rivera, Nora A. Fierro, Ana Olivera, and Ryo Suzuki

B Cells and Autoantibodies in the Pathogenesis of Multiple Sclerosis and Related Inflammatory Demyelinating Diseases
Katherine A. McLaughlin and Kai W. Wucherpfennig

Human B Cell Subsets
Stephen M. Jackson, Patrick C. Wilson, Judith A. James, and J. Donald Capra

Index

Volume 99

Cis-Regulatory Elements and Epigenetic Changes Control Genomic Rearrangements of the IgH Locus
Thomas Perlot and Frederick W. Alt

DNA-PK: The Means to Justify the Ends?
Katheryn Meek, Van Dang, and Susan P. Lees-Miller

Thymic Microenvironments for T-Cell Repertoire Formation
Takeshi Nitta, Shigeo Murata, Tomoo Ueno, Keiji Tanaka, and Yousuke Takahama

Pathogenesis of Myocarditis and Dilated Cardiomyopathy
Daniela Cihakova and Noel R. Rose

Emergence of the Th17 Pathway and Its Role in Host Defense
Darrell B. O'Quinn, Matthew T. Palmer, Yun Kyung Lee, and Casey T. Weaver

Peptides Presented *In Vivo* by HLA-DR in Thyroid Autoimmunity
Laia Muixí, Iñaki Alvarez, and Dolores Jaraquemada

Index

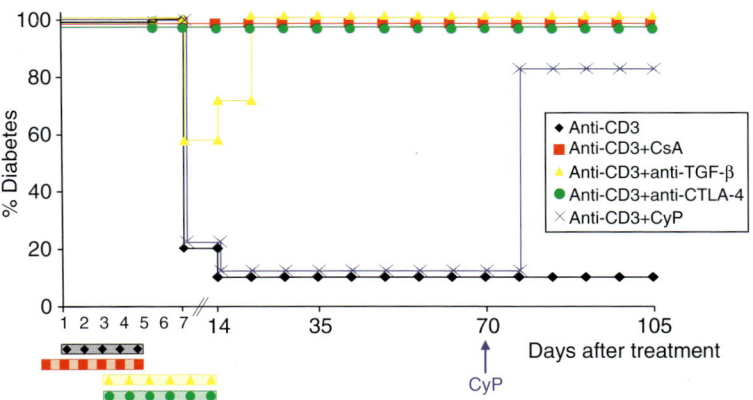

PLATE 1

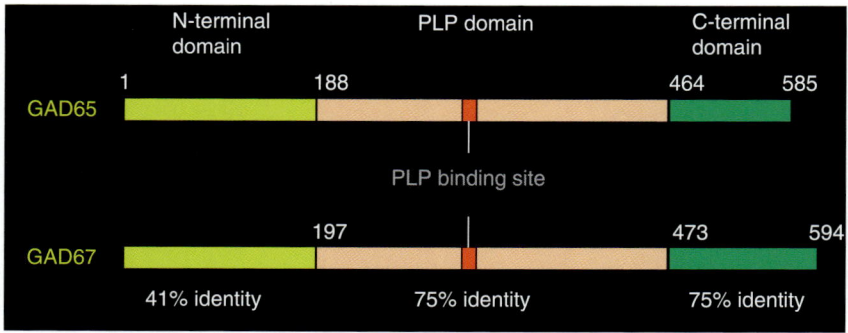

PLATE 2

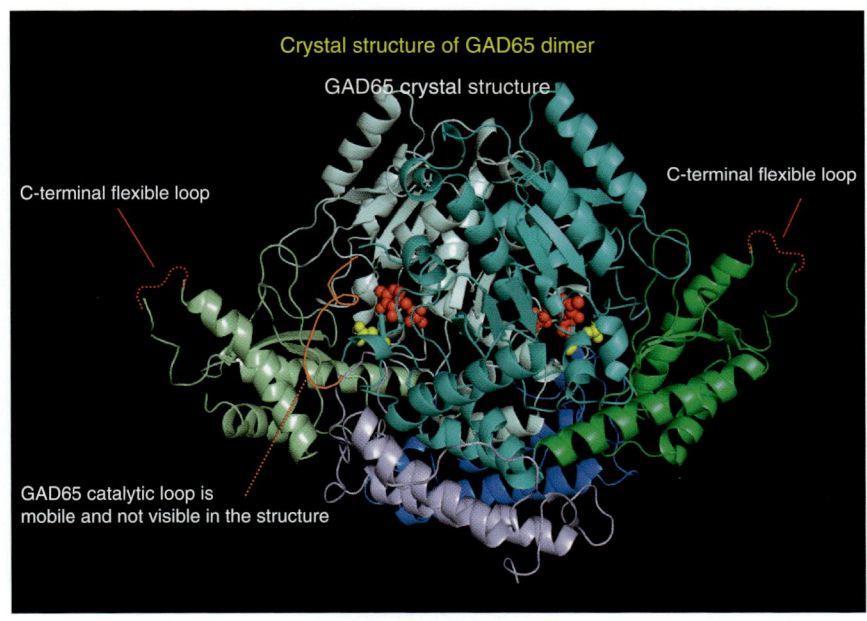

PLATE 3

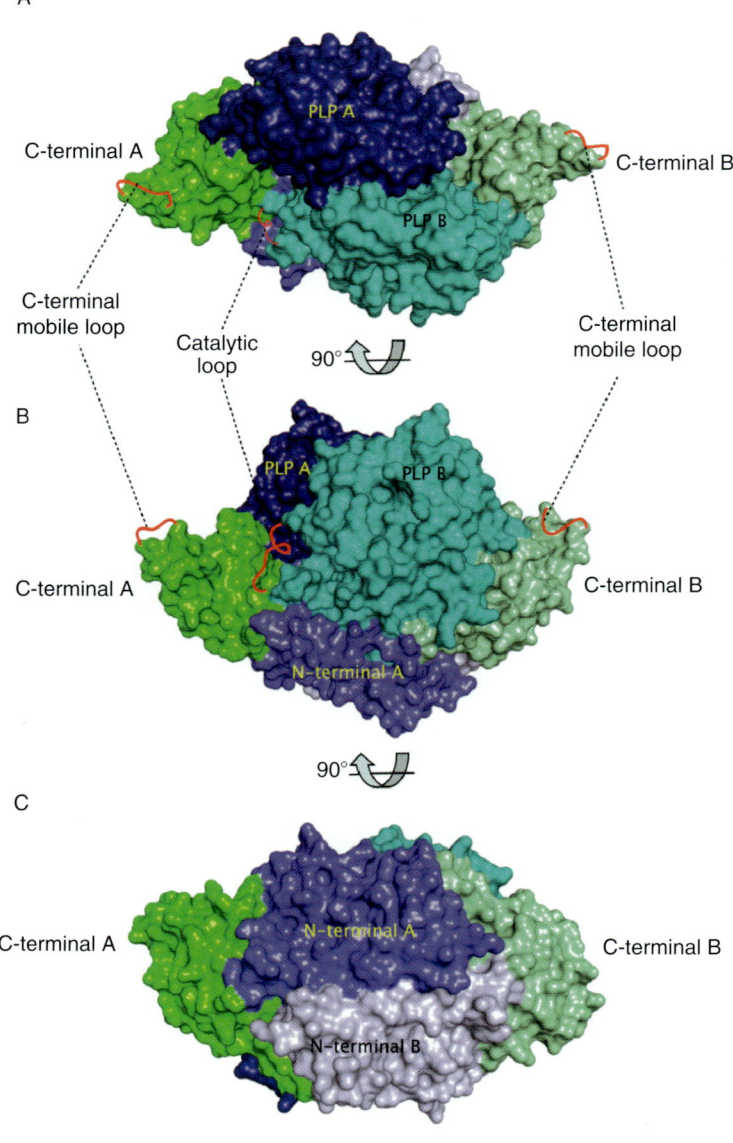

PLATE 4

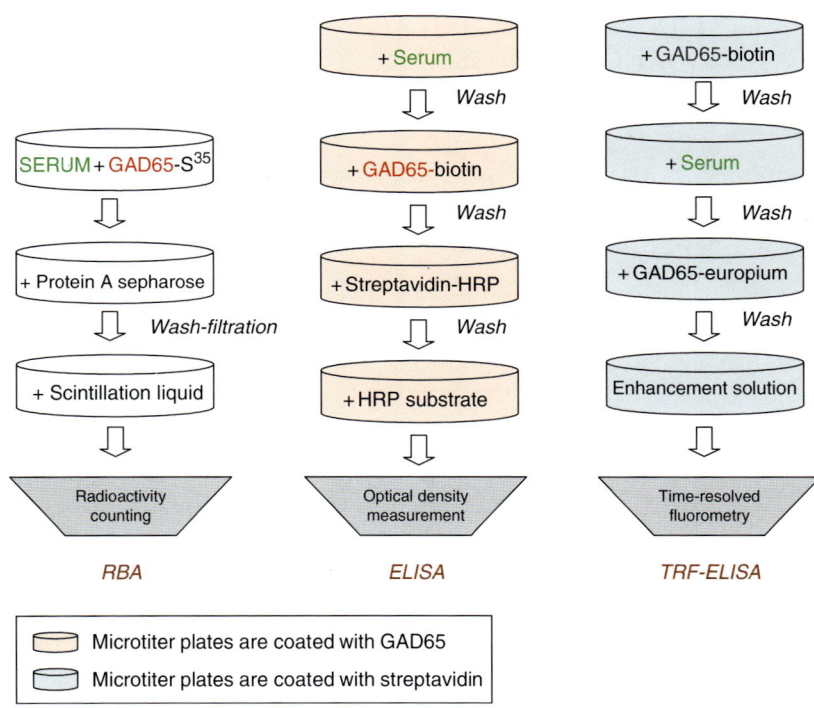

PLATE 5

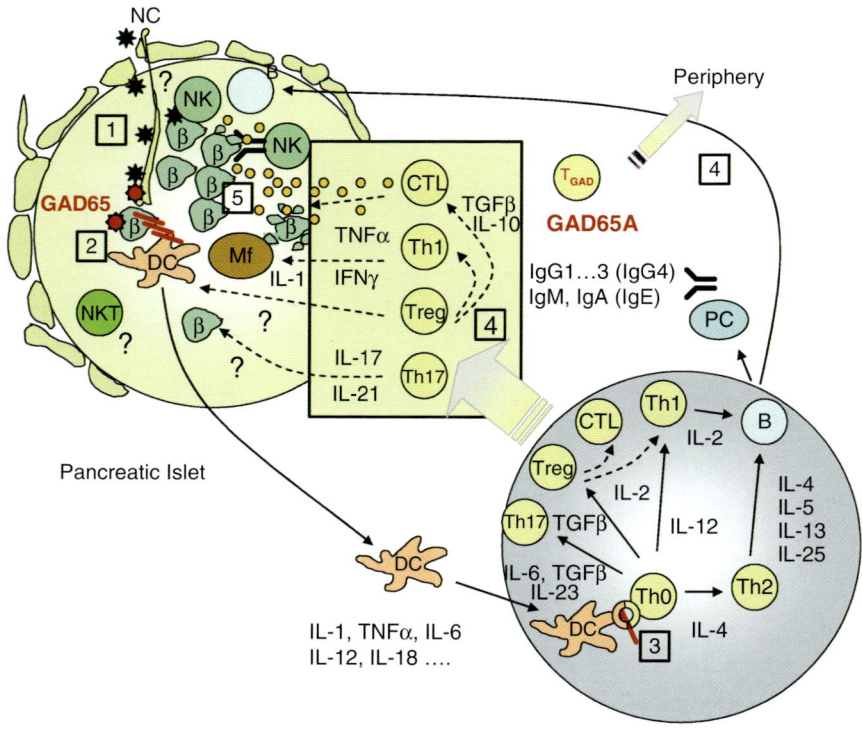

PLATE 6

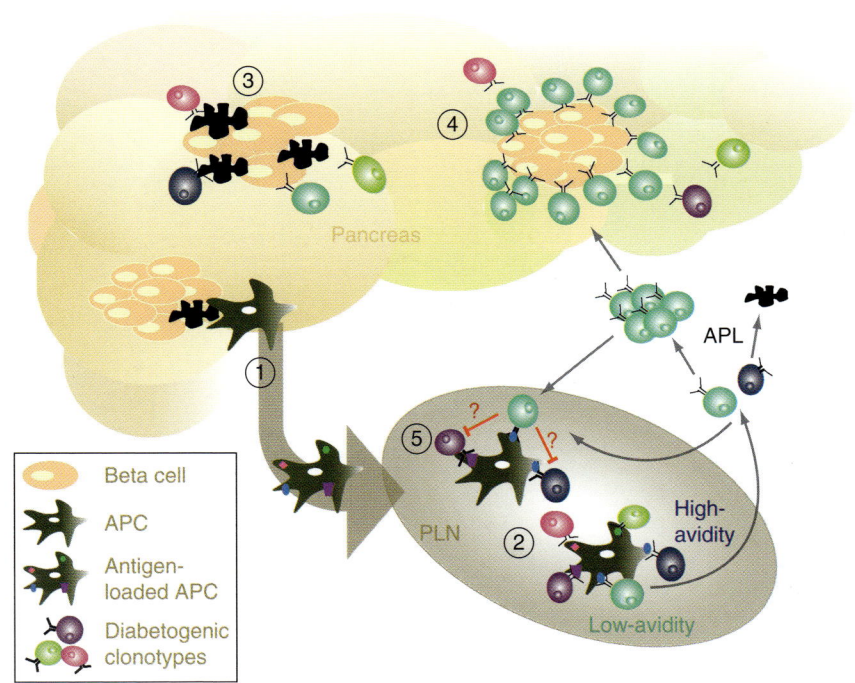

PLATE 7